D0408109

WITHDRAWN
UTSA Libraries

Adapting to Climate Change

Adapting to Climate Change

Markets and the
Management of an
Uncertain Future

Matthew E. Kahn

Yale UNIVERSITY PRESS
New Haven and London

Library
University of Texas
at San Antonio

Published with assistance from the Mary Cady Tew Memorial Fund.

Copyright © 2021 by Matthew E. Kahn.
All rights reserved.
This book may not be reproduced, in whole or in part, including illustrations, in any form (beyond that copying permitted by Sections 107 and 108 of the US Copyright Law and except by reviewers for the public press), without written permission from the publishers.

Yale University Press books may be purchased in quantity for educational, business, or promotional use. For information, please e-mail sales.press@yale.edu (US office) or sales@yaleup.co.uk (UK office).

Set in Galliard type by IDS Infotech Ltd., Chandigarh, India.
Printed in the United States of America.

Library of Congress Control Number: 2020944842
ISBN 978-0-300-24671-1 (hardcover : alk. paper)

A catalogue record for this book is available from the British Library.

This paper meets the requirements of ANSI/NISO Z39.48-1992 (Permanence of Paper).

10 9 8 7 6 5 4 3 2 1

**Library
University of Texas
at San Antonio**

Contents

Introduction: Why Adaptation?

In 1980, the biologist Paul Ehrlich and the economist Julian Simon engaged in a famous debate. Ehrlich argued that ongoing population growth would lead to overconsumption of natural resources and a collapse in food consumption per person. Simon countered that rising scarcity creates incentives to unleash human ingenuity and address the challenge.

Flash forward to the year 2020, when we were confronted by a risk that again posed existential questions. The covid-19 crisis was a wakeup call against complacency about our standard of living. The economic dislocation and disruption of daily life caused by social distancing and other measures taken to reduce the likelihood of infection rocked the global economy to its foundations. A silver lining was that we learned how quickly the world can adjust to shocks that we did not anticipate and for which we were unprepared. But it remains an open question how nimble we will be in responding to future threats.

During the covid-19 pandemic, the challenge of climate change was temporarily displaced from the news. But the challenge has not gone away. This book revisits Julian Simon's core hypothesis that human ingenuity plays a central role in reducing climate risk because our ability to adapt is accelerating over time.

Coping with Climate Change Risks

Climate shocks affect people both rich and poor. In December 2017, several fires raging in Southern California destroyed homes and polluted the air. After the fires were put out, several hillside homes in Montecito were wiped out by mudslides triggered by heavy rain falling on bare slopes. The next year, celebrities including Miley Cyrus and Robin Thicke lost homes in the Woolsey Fire in Southern California.[1] At the same time, more than a hundred homes burned down in the Seminole Springs mobile home park off Mulholland Highway. "There's a lot of poor people that lost houses there," resident manager Doug Olson told the *Los Angeles Times*. "It's a total tragedy."[2]

We have two chief strategies to cope with the threat of climate change. First, we can mitigate the threat by reducing our production of greenhouse gas emissions: mitigation represents a type of insurance policy that reduces the likelihood of the most dangerous climate change scenarios. Second, we can adapt to the threat by changing how we live in response to new threats.

This book focuses on the second strategy. It explores how people, firms, and governments can adapt to the new risks posed by climate change, how we can change our lives even as the climate change challenge grows more severe.

Given what we know, the rational strategy in the face of considerable and rising climate risk is to engage in both mitigation and adaptation. But at least in the short term, global greenhouse gas emissions will continue to rise as world per capita income increases. For the billions of people in the developing world who seek a better life, these income increases will be a good thing. However, given current technologies, continued emissions will exacerbate the climate change challenge, which will make adaptation even more important to guarantee future improvements in people's standard of living.

The Mitigation Challenge

Reducing greenhouse gas emissions will require burning fewer fossil fuels (coal and natural gas) to generate electricity, heat and cool our homes and workplaces, and power transportation services. Yet the global use of fossil fuels is continuing a steady upward increase. In the developing world, energy consumption has risen sharply in recent decades.[3] In China, India, and other growing developing nations, the demand for air conditioning, private transport, and energy to power manufacturing is producing ever more greenhouse gas emissions. If China achieves the US rate of car ownership by the year 2070, roughly one billion vehicles will be on its roads. If these vehicles use gasoline as fuel or if they use electricity generated by fossil-fueled power plants, the transportation sector will continue to be responsible for very large and growing quantities of greenhouse gas emissions.

In the absence of a global tax on fossil fuels reflecting their environmental consequences, each consumer of fossil fuels ignores the social costs imposed on others by fossil fuel use. Given that the atmosphere is common property, the economic challenge known as the tragedy of the commons arises. Every nation hopes that every other nation will take costly action to reduce its own emissions. In the face of this free-rider problem, mitigating the climate change challenge has relied on international treaty negotiations. These diplomatic efforts have led to major international events culminating in the Paris Climate Conference in December 2015, which attracted 190 nations and resulted in the Paris Agreement on climate change.[4] However, although this conference received widespread media coverage and praise, the agreement lacked clear enforcement rules. Several major nations, including China and India, committed to reduce their carbon intensity (greenhouse gas emissions per dollar of gross national product, or GNP), not their actual level of carbon emissions. A nation whose greenhouse gas emissions increase by 20 percent while its national income increases by 40 percent will experience a reduction in its carbon intensity at the same time that it increases its total emissions. Given these limitations, in August 2018 the *Economist* published a lead story headlined, "The world is losing the war against climate change." "Optimists say that decarbonisation is within reach," the article observed. "Yet, even allowing for the familiar complexities of agreeing on and enforcing global targets, it is proving extraordinary difficult."[5]

How high global carbon dioxide concentrations will go remains an open question. In April 1958, the global atmospheric carbon dioxide level was 317 parts per million. Over the past sixty years, this level has risen linearly such that, by April 2020, it stood at 416 parts per million. The global carbon dioxide level will likely rise over the next decades due to increased fossil fuel consumption in the developing world.

In the United States and Western Europe, many young people are deeply concerned about the climate change challenge. Inspired by such environmental activists as Greta Thunberg, many youth seek to hold elected officials accountable for their slow carbon mitigation efforts. If new generations of young people continue to prioritize the climate change challenge, a political shift could occur over time as more and more voters in developed nations emphasize the climate challenge.

The diffusion of electric vehicles, especially if significant progress in battery storage technology were made, could alter the steady rise in atmospheric carbon dioxide. Similarly, an increase in the percentage of power generating by clean, renewable sources such as solar, hydro, and wind would reduce the world economy's carbon intensity (greenhouse gas emissions per dollar of GNP). Economists broadly agree that the introduction of a carbon tax would accelerate this energy transition.[6]

Despite efforts by the progressive movement and by young people, however, few governments around the world have implemented a carbon tax. Nor have economists convinced elected officials to enact a cap and dividend program that taxes greenhouse gas emissions and then returns the revenue collected to those who consume the most fossil fuels. The challenge is to design a policy proposal that simultaneously puts a price on carbon emissions without lowering the real incomes of those who work in the fossil fuel sector, own assets whose value is tied to fossil fuel use (such as Houston real estate), or are reliant on fossil fuels to live their lives. Such individuals will lose in the short run from introducing carbon pricing incentives, and so far they have fought the introduction of such carbon taxes.

The yellow vest protests in France have highlighted that rural people rely heavily on fossil fuels for transport and for heating services and will oppose policies that will raise the energy prices they pay. Similarly, workers whose income is tied to the extraction and distribution of fossil fuels, such as coal and oil workers, have a financial incentive to oppose carbon pricing, as do the owners of shares of stock in energy companies. In general, having a financial stake in the fossil fuels economy creates interest groups that seek to block new policies to mitigate climate change. Facing this reality, progressives in the US Congress have not recently proposed the introduction of carbon pricing. Instead, they have focused on promoting a low-carbon economy through such regulations as increasing vehicle fuel economy. Individual states such as California, for example, are sharply boosting the legislated share of electricity that must be generated by renewable power.

Developing countries continue to prioritize their own economic growth over embracing and enforcing a global climate treaty that would cap global emissions. If developing countries' emissions continue to rise, adaptation to climate change will be a central issue on the global agenda for decades. Thus, we need to understand how people, firms, and governments interact through markets to reduce the risk posed by climate change.[7]

Climate Change Adaptation: The Microeconomic Perspective

The microeconomic approach to studying climate change adaptation emphasizes the role that markets and human capital play in coping with new risks. Markets facilitate trade among diverse individuals. The value of such trades increases when people and firms face more severe shocks. When people know that they can participate in markets, they have an incentive to develop skills that are rewarded by the market. Market prices send signals regarding what talents, goods, and services are scarce.

This book focuses on the role that a variety of markets, including markets for real estate, labor, capital, food, and insurance, play in fueling adaptation. In the chapter on farming, for example, I discuss how collection of big data by farmers is allowing them to better calibrate their responses to weather shocks so that the quantity and quality of their produce suffers less during bad weather events. As this example demonstrates, we are not passive victims in the face of the punches Mother Nature throws at us. If we can anticipate or at least partially predict these shocks, we can be better prepared and suffer less when shocks such as drought and heat waves actually occur.

Given that each of us differs with respect to our goals, talents, and knowledge, we all have different capacities to cope with change. The microeconomic approach to studying climate adaptation recognizes this point.

Whereas macroeconomists often focus on the so-called average person, microeconomists are explicitly interested in distributional effects. We focus on how different segments of the population cope with emerging threats. At a time of growing concern about income inequality, it is essential to understand how different groups cope with new risks.

Richer people have access to a greater set of products to help them cope. Younger people are generally more flexible because they have not built up a stock of lifetime experience and have not yet planted roots. Investment in education helps individuals build up human capital. Their enhanced skill and knowledge then facilitate problem solving and expand their imagination in handling situations they have not seen before.

An equitable and fair society recognizes that some segments of society have a limited capacity to cope with change. Examples include people who are unaware of the challenge, who cannot afford products that facilitate adaptation, who are deeply tied to their current location and community, and who do not want to change their day-to-day life.

Academic research, experimentation with different coping strategies, and the accumulation of human capital each play a key role in helping people adapt to new risks. Just as amateurs can learn how to play chess or to drive a car, building up the skill to adapt to new challenges is an investment. It requires incurring costs, but it also offers benefits in terms of reducing one's exposure during a time of increasing climate risks.

In 2018, William Nordhaus was awarded the Nobel Prize in Economics for his pioneering work on the economics of climate change. This book builds on his work. Nordhaus pioneered the use of mathematical models to study the interconnections between macroeconomic growth and climate change. Using a macroeconomic model to explore the costs and benefits of a carbon tax, he sought to be precise about the tradeoffs among different policy options. He argued that economic growth exacerbates the challenge of climate change, while climate change, by raising the world's average temperature level, acts to slow economic growth. By using a system of equations to study the feedback between world economic growth and climate change, Nordhaus sought to identify the optimal tax policy for mitigating the climate change challenge.

A key equation in Nordhaus's model is the damage equation, which expresses the relation between climate change and economic damage (often measured as the loss in the average person's consumption).[8] To simplify the mathematics in his core model, he assumes that the cause (climate change) always has the same effect. Such a model effectively assumes that adaptation does not occur.

Building on Nordhaus's work, I argue that the damage caused by climate change is falling over time due to adaptation. Even though we are confronting greater overall levels of greenhouse gases in the atmosphere, people, firms, and governments are becoming better and better at adapting to the challenge that we have unleashed because of the evolving microeconomy.

This bold claim is being widely tested by many independent scholars doing research in the field of environmental economics. This burgeoning field has enjoyed a great decade as many young scholars have used big data to explore how climate conditions affect our economy's productivity and our quality of life. Much of this research has used creative research designs to study climate cause and effect statistical relationships. This book looks back and synthesizes insights from this field.

In 2010, I published *Climatopolis*, in which I discussed the future of US cities in the face of climate change. In that book I argued that cities prosper if the skilled choose to live and work there. Cities with a declining quality of life suffer a brain drain, become poorer, and lose their tax base. This competition to grow, attract, and retain the skilled gives cities an incentive to invest in climate resilience. *Adapting to Climate Change* delves further into this theme.

Over the past decade environmental economists have made great progress in using new geocoded data to study claims such as those I advanced in *Climatopolis*. For example, consider the familiar PSAT tests that college-bound high school juniors take. Researchers documented that students who take the PSAT test after being exposed to more hot days over the previous year score lower on this standardized exam.[9] This is a creative example of estimating the climate damage function. Students learn less when they are exposed to high heat in the classroom.

This finding is not an immutable law of physics. The researchers find that the negative effects of heat on learning vanish for schools that have air conditioning. Given that public schools in poor school districts are less likely to have central air conditioning, this result has important equity implications. In this specific case, the adaptation hypothesis posits that if a public school were funded so that air conditioning could be installed, the weather penalty would vanish. Research that documents the costs for disadvantaged students of not having air conditioning in their schools increases the likelihood that such infrastructure will be installed. In 2020, the city of Baltimore invested millions of dollars in just such a program. Once the air conditioning is working, the optimistic adaptation hypothesis posits that student learning will continue to take place even in schools located in places that are experiencing high recent heat.

This example highlights a major theme of this book. Households, firms, and governments, with the help of quantitative academics, can use big data to identify challenges resulting from a changing climate. Such diagnosis of challenges is a necessary but not a sufficient condition for accelerating adaptation. Armed with this knowledge, decision makers have strong incentives to seek out cost-effective solutions. In this way, iteratively discovering what ails the patient and then looking for a cure fuels adaptation.

Given that our economy is constantly changing as we obtain more human capital, experiment, and create new markets, no immutable cause-and-effect relation exists between climate conditions and economic outcomes. Rather, how we configure our society determines how climate change affects our quality of life. Over time, we have an ever-increasing capacity to adapt to more climatic effects. Although it can be costly to adapt to the new risks, these costs are continually declining over time.

This optimistic view dovetails with the research of Paul Romer, the other 2018 Nobel Prize winner with William Nordhaus. His research on endogenous technological change emphasizes the importance of knowledge accumulation and discovery in market economies. Once an idea has been discovered (think of the creation of Microsoft Word), others can adopt it at low cost, thereby boosting their productivity.

The rising awareness about the risks posed by climate change creates a growing market for adaptations, ranging from construction materials that can withstand natural disasters to more powerful air conditioners. In this sense, threats to people's health and quality of life and to firms' profitability create new opportunities for those who can devise solutions. The defining feature of our modern economy is the continual introduction of new goods—think of Google, the cell phone, or Uber. I build on Romer's ideas from the perspective of microeconomics, arguing that the economy is building resiliency through a series of small and subtle adjustments that play out each day.

The market for air conditioning is an example. As outdoor temperatures rise, people will increasingly demand more powerful air conditioning. This aggregate demand creates a market for firms to design products to meet this demand. Such mass production leads to economies of scale, which lowers the cost per unit produced. As a result, lower-income people can afford these goods. As lower-income people purchase such adaptation-friendly products, they suffer less from extreme weather events. Academics using big data can rigorously test whether the damage caused by such extreme weather is shrinking for poorer people. If this is the case, then this is direct evidence that capitalism is playing a productive role in protecting the most vulnerable people from emerging threats.

The Challenge of Integrating Climate Science and Climate Economics

This is not a book about climate science. I take it for granted that the climate science community will continue to make progress in refining climate prediction models. Such models seek to answer a variety of questions with important implications for the economy. For example, a climate model might be used to predict how many 90-degree Fahrenheit or hotter days Berkeley, California, will experience in summer in the year 2070.

Berkeley summer heat in the year 2070 represents a random variable today. As we think ahead, we can form our best guess about its range of likely outcomes. An ever-growing literature in the social sciences, incorporating economics, psychology, and sociology, focuses on how and when people update their beliefs about the world and the risks we face. Very few of us anticipated the covid-19 pandemic of 2020. Going forward, will we now overestimate the risk of future pandemics, or will we become complacent with the potentially false belief that the challenge will not arise again?

The same issues arise in forming expectations of emerging climate risks. Climate scientists are learning about the new risks even as they are educating the public about these risks. During an era of political and income polarization, the credibility of experts has been questioned and the troubling challenge of fake news has emerged. If large proportions of the public ignore what environmental scientists say, and if the scientists' predictions are accurate, a democratic society may underinvest in self-protection against emerging threats.

Adaptation could be accelerated if people, firms, and governments keep up with the experts' views. In debates over how people update their beliefs about the future, two Nobel laureates have proposed a humble approach for studying how people make choices when faced with fundamental uncertainty.[10] Their starting point is that people are risk averse and are aware that they do not know exactly how the world operates (especially at a time when these risks are changing). This theory posits that individuals make their choices by following robust decision rules. These rules focus on making the best possible choice when faced with the worst-case scenario. Such humble decision makers will invest more time and effort in protecting themselves from emerging risks because they know that they do not know the new risks they are facing.

The emerging field of behavioral economics offers an alternative narrative for how we assess emerging risks. Behavioral economists posit that, in forming expectations about future risks, people often make mistakes as they focus on salient information (such as high-profile recent events) and ignore information that challenges their worldview. Such individuals will be more likely to buy real estate assets that are at risk. Such buyers may be blissfully unaware about the risks they have exposed themselves to. The behavioral economics school of thought is more pessimistic about our ability to adapt to new risks. Climate change represents a high-stakes setting for testing the ideas embodied in behavioral economics and neoclassical economics.[11]

Household Adaptation to Emerging Climate Risks

Parents seek to rear happy, safe, and comfortable children who will grow up to be functioning adults. A household produces these goods by using its own time and buying market inputs such as living in a safe area, installing an alarm system, or allocating time to a purposeful activity (perhaps a self-defense class).

Climate change poses new risks to a family's health and quality of life. A family that is aware of these risks has an ever-growing variety of adaptation strategies to counter these risks, ranging from living in an easier-to-cool structure to migrating to a safer area. Richer households will have access to more strategies than poorer households.

No one individual's demand (not even Bill Gates) drives corporate research and development patterns. However, aggregate demand for new products directs innovation. Each year innovators launch new products, such as Diet Coke, Tesla automobiles, or the iPhone. These inventions are examples of directed technological change. Innovators focus on innovations that are expected to be demanded and thus profitable. Over time, due to declining costs of production (largely resulting from global supply chains), more people can afford them. Consider air conditioning again. Over time its quality-adjusted price has declined, which has allowed more and more people to afford it. By operating an air conditioner, a household can reduce its exposure to extreme heat, reducing the damage caused by that heat and improving quality of life. This simple example highlights how the climate damage function shifts depending on the actions taken by individuals.

The microeconomic approach for studying climate change adaptation explores who has an edge in coping with change and how one builds this capacity. For example, those who love a specific city and a specific leisure activity within that city will be less nimble in coping with a city-specific climate shock because they will be less willing to move elsewhere. Someone who loves skiing and is unwilling to substitute a different leisure activity will lose more than other people if local temperatures in winter often rise above freezing. The same issue of exhibiting a willingness to substitute arises in evaluating the impact of climate change. Consider your diet. Suppose you enjoy eating strawberries and prunes equally, but climate change raises the price of strawberries (perhaps because of the water needed to grow them). In this case, you will not suffer much because you can substitute prunes for strawberries. This key economic idea of substitution arises again and again. Those who are unwilling or unable to substitute will suffer more as climate change disrupts past routines and consumption habits.

Income inequality and persistent poverty continue to be key social issues. Research documents a stubborn and growing persistence of earnings. The wealthy's children are increasingly likely to be rich while the poor's children are increasingly likely to be poor.[12]

Climate change exacerbates this inequality challenge. The poor have the least ability to afford products to help them offset climate risks. This group lives in the riskiest places, in the lowest-quality housing, has the highest-risk jobs, and has the least access to health care and other inputs. Chapter 3 examines how poor people and poor places are likely to be affected by climate change.

The Public Infrastructure Challenge

When extreme climate events and natural disasters strike, public infrastructure is often overwhelmed. On a typical day in a major city, people travel by plane, train, and automobile for work, pleasure, and shopping. People rely on the electricity grid to provide a constant stream of services for households and firms. Clean water and proper sewage treatment are essential to daily life.

The urban grid represents a multibillion-dollar investment that is costly and inconvenient to upgrade once it is built. Yet much of our durable infrastructure has been built in places and in ways that now expose it to risk. Civil engineers have the expertise to diagnose these risks and to suggest investments to upgrade them, but these improvements can be costly. Given that capital upgrade costs are borne now but resilience benefits are gained in the future, elected officials with a finite horizon (for example, because they face term limits) may not have the right incentives. For example, they may be unwilling to invest tax dollars in projects that will provide protection only when extreme climate events occur, such as extreme rainfall events.

Remaining Productive in a Hotter World

In choosing where to locate its HQ2, Amazon was lured by many major cities. This high-profile decision demonstrates how footloose firms choose where to locate as cities compete for economic advantage. San Francisco's average salaries are high because it has attracted and retained such firms. One theme of my research over the past twenty years has been that cities with a high quality of life attract skilled people and the firms who hire those people. If a city is becoming unlivable, major employers will leave or choose not to enter. Low quality of life is an anti–economic development strategy. City leaders therefore have strong incentives to upgrade their cities so that they remain competitive in the market for productive firms.

Empirical research using data at the national, firm, and individual levels has demonstrated that we are less productive when we are exposed to extreme climate conditions such as heat or natural disasters. This research has direct implications for the future productivity of the US economy.

If a firm's managers are aware that climate conditions affect a firm's productivity, they can take steps to protect the firm against shocks. Adaptation strategies include moving the firm to a safer location, upgrading a production facility, or shifting a work schedule to avoid high afternoon temperatures. Each strategy features costs and offers benefits. During the economic disruption caused by the covid-19 pandemic, many firms gained valuable lessons in keeping workers productive even as they telecommuted. The lessons learned from this crisis represent valuable human capital that will lower the cost of adapting to future natural disasters.

A publicly traded firm's stock prices will decline if expected profits are falling because the firm is not proactively addressing new climate challenges. In this case, shareholders will suffer an asset value loss and will lobby the firm's leadership to act. This example of self-interest fostering resilience highlights how capitalism creates vested interest groups with a stake in promoting adaptation.

While Amazon, Google, and other high-tech companies are highly profitable and treat their skilled workers well, tens of millions of American workers are not employed in the tech sector. In chapter 5, I explore how other firms are likely to perform when faced with new climate threats. For example, which firms have an edge in adapting to emerging risks: small firms or big firms, firms with weak managers or firms with strong managers?

The vast majority of the US economy may now be centered in cities and their nearby suburbs, but millions of people still work outside cities in the resource and extraction sector. Chapter 5 also discusses how the productivity of specialized single-industry towns (such as fishing areas) may be affected by climate change. Such places have made a risky bet by specializing in one economic sector.

Real Estate Markets and New Climate Risks

Climate science modelers have predicted the geography of emerging risks from temperature extremes, sea level rise, heat, and natural disasters. As the fires and mudslides in California in 2017 and 2018 demonstrated, even paradise can suffer. The popular media routinely publish articles discussing the rising flood risk that parts of Miami now face. The real estate database company Zillow has estimated that perhaps $800 billion of US coastal real estate will be at risk due to sea level rise by the year 2100.[13]

The price of real estate varies across cities and across neighborhoods. A property's location determines the occupant's local labor market opportunities, the length of one's commute, and a family's day-to-day access to amenities. Areas with a high quality of life will command a real estate premium. Chapter 6 explores how the housing market will cope with emerging climate risks in cities. Since 64 percent of US adults are homeowners, most of us have made a bet that the specific city and neighborhood in which we live will have a bright future. Climate risks such as extreme heat, flooding, and fire could lower housing wealth for property owners by lowering the market price for real estate.

A home is the most valuable asset of most households. This asset's value is crucial for determining a household's wealth and, especially in retirement, their consumption. The possibility that one's home could sharply fall in value because of climate shocks creates strong incentives for people to consider risks, whether they are buying a new home or taking steps to reduce their home's risk exposure. This demand for real estate resilience creates an incentive for firms that can supply market products that facilitate resilience (that is, protect homes from fires and floods). Whether innovations can be effective in protecting real estate from emerging climate risks will be a key determinant of the future cost of climate change.

Accelerating Adaptation Progress

In the first half of this book I explore how different sectors of the economy are affected by climate change. In the second half I pivot to the future and explore emerging trends that facilitate adaptation. Economists celebrate the evolutionary nature of capitalism as market prices signal what goods are relatively scarce. The ever-increasing quantity and quality of big data help us learn in real time about the challenges we now face.

How the big data revolution facilitates adaptation is the subject of chapter 7. In December 2017, the Thomas Fire in Los Angeles came within four miles of my house. On my Twitter feed, I received real-time updates about the fire's path and neighborhood evacuations. This information reduced my anxiety and helped me plan. Similarly, the big data revolution helps for-profit firms like Amazon anticipate demand for specific products. This predictability facilitates inventory accumulation and the logistics of next-day deliveries to impatient consumers. The big data revolution also fuels government responsiveness to citizen concerns as people post information about the real-time challenges they face and governments experiment with different services to address their concerns.

Chapter 8 explores how encouraging more renting (and less home ownership) will facilitate adaptation. Even though home ownership has been a key part of the American Dream for decades, it represents an increasingly risky bet as households choose to invest most of their assets in a single location that faces increased climate change risk. Encouraging more young people to rent will open up several valuable adaptation pathways. Such footloose renters can more easily move as they learn about emerging place-based risks. Renting allows people to hold a more diversified asset portfolio, thus shielding their wealth from place-based shocks such as natural disasters. A phase-out of the mortgage interest deduction and requiring homeowners to pay taxes on imputed rent will remove some of the subsidies that encourage home ownership today.

Adaptation faces policy roadblocks that inhibit the reallocation of scarce resources (such as land) to its highest value use as the climate changes. In chapter 9, I propose new land use rules and urban transport rules that would greatly accelerate the process of adaptation. Consider the speed with which someone can move through a city. This is a function of transportation policies, such as road conditions, investments in public transit, and the regulation of automated vehicles. If you can move at higher speeds, living near jobs and cultural opportunities becomes less valuable to you. In an extreme example of moving at very high speeds for relatively little cost, you could live in Anchorage and work in New York City. The ability to move at higher speeds increases the menu of alternative locations where people can live, which could help them adapt to a wide range of risks.

Research in climate economics argues that the farming sector has been slow to adapt to climate change.[14] Food production takes place outside and thus faces greater exposure to climate change than other sectors of the economy. Research also documents that agricultural yields can be quite sensitive to extreme heat and drought conditions, suggesting that this sector's productivity could be greatly affected by climate change. In chapter 10, I discuss potential adaptation pathways for the agricultural sector. Possibilities include moving some farming activity indoors, migrating farm activity to different locations, and using big data and real-time monitoring to track the productivity of different plots of land.

Globalization represents the international flow of labor, goods, and capital across borders. The Trump administration has emphasized the costs of immigration. The ability to migrate across space represents an important adaptation strategy. It also represents a key investment decision as families seek out destinations offering a better quality of life. In recent years, large numbers of international refugees have sought to move to safer, richer areas. Such migration is a key adaptation strategy for those who are suffering in their country of origin. But when large numbers of refugees from different cultures enter a nation, migration can trigger resistance. Several nations have recently erected international migration walls. In Chapter 11, I discuss how to design international rules to facilitate the gains to trade between those who wish to move across international borders and nations willing to admit new immigrants. Adaptation to new climate risks will be facilitated if such international migration barriers decline. Migration scholars emphasize that international migration offers "gains to trade." Michael Clemens, for example, posits that world gross domestic product could double if there were open borders.[15] In addition to discussing international trade in labor, I consider how globalized markets in food and capital facilitate adaptation. Walls and barriers to international trade hinder adaptation.

The microeconomic approach to studying climate change adaptation focuses on individual choice, the role of market competition, and the role of human capital. People choose where they live, how they live, what labor markets they participate in, and what assets and products they buy. Such individuals have strong incentives to consider how these choices affect their family's health and safety. As the risk of climate change becomes salient, households are more likely to take actions to protect themselves, and market product suppliers will find profitable opportunities delivering such products.

Competition guarantees that consumers have market choices and access to new products developed and marketed by companies that anticipate a new market niche. If farmers are unable to adapt to changing weather conditions, they can sell their lands to other farmers who can make better use of this asset. If a publicly traded company underperforms in the face of climate change because it fails to invest in resilience and suffers from place-based shocks, its stock price will decline and leveraged buyout firms may purchase shares and replace the incumbent management. This competitive threat in the asset market encourages the current leadership team of a company to stay awake and scan the horizon for emerging threats.

Throughout this book, I emphasize the role that human capital plays in facilitating our ability to engage in problem solving. Human capital not only raises one's earnings but also helps people be healthier and raise children who can adapt to continued change. Our global stock of human capital is rising as people in the developing world invest more in their own education.

The Great Race

A great race is unfolding as global greenhouse gas emissions continue to rise but we become better at adapting to the emerging threat. In this book, I highlight what academic environmental economists have learned about how people, firms, and governments cope with new risks. These studies offer crucial insights into how much we could suffer in the future if we fail to adapt.

In 1980, Julian Simon said that "discoveries, like resources, may well be infinite; the more we discover, the more we are able to discover." Each chapter of this book reflects Simon's optimistic theme of the role that human capital and ingenuity play in building our individual and collective resilience.

1

A Microeconomics Perspective on Climate Science Prediction

We are uncertain about how climate change will affect our future quality of life. Few of us are good at predicting the future. Future events such as which team will win the Super Bowl in the year 2030 or who will be elected president in 2032 are random variables today. The New England Patriots could win the Super Bowl in the year 2030, but they may not. Given that few of the current players will still be on that team in the year 2030 and the team will likely have a new coach, it is quite challenging to predict this future probability given what we know today.

Consider two different types of random variables: presidential election outcomes and games of cards. Back in 2015, the outcome of the 2016 presidential election was not known. At that time, few experts predicted that Donald J. Trump would win the election. Some experts thought he had no chance of winning.

Contrast this prediction problem with the problem faced by gamblers in Las Vegas. Experienced Las Vegas blackjack players can estimate the probability that the sum of their cards will exceed twenty-one if they ask for another card when their current cards add up to seventeen. Such players can rely on past experience and what they have learned from friends and the Internet to make educated predictions concerning whether the next card (whose value is a random variable) equals four or less.

Unlike the situation of playing blackjack, rising global greenhouse gas emissions means that our past experience with the weather is less informative about future risks. Even though Berkeley, California, has never experienced a 105-degree day in June over the past hundred years, it would be a mistake to extrapolate from this past long time series to predict that the probability is zero that Berkeley will experience such a hot day in the year 2040. In this sense, climate scientists are researching a moving target. The public is asking these experts for their best guesses about what the future holds. If these experts are honest about their own uncertainty about the future, does the public dismiss them as fools or thank them for their modesty? In dealing with new threats, we face great danger if we trust experts whose forecasts are based on incorrect models.

Our ability to adapt to changing conditions is hindered if we are unaware that climate conditions are changing. People are less able to cope with a new risk if they believe that the risk is unlikely to occur. The covid-19 pandemic is imposing great costs on every nation around the world because we were so unprepared for this shock.[1] The conventional wisdom is that, for many nations, the Spanish Flu of 1918 occurred so long ago that people downplayed the risk. In contrast, nations in Asia such as Taiwan had recent experience with the SARS epidemic of 2003 and were thus better prepared for handling the pandemic.

This chapter explores the interplay between climate scientists and different members of the public, firms, and governments.

Forecasting Progress

Progress in computing power and climate science means that short-run weather forecasts are growing increasingly accurate. Writing in *Nature,* researchers assert: "Scientific and technological developments have led to increasing weather forecast skill over the past 40 years. For example, forecast skill in the range from 3 to 10 days ahead has been increasing by about 1 day per decade. Today's 6 day forecast is as accurate as the 5 day forecast ten years ago."[2] In the case of tornado prediction, new climate models are providing more precise geography and are available sooner. For example, tornado severity can now be predicted and disseminated eighteen minutes before a tornado hits. This increased window gives people more time to seek shelter.[3]

Climate models are evaluated based on their ability to predict past temperature dynamics and flooding events. We already know whether a specific place suffered a severe heat wave or flood in the years 2000 to 2015. Climate models can be benchmarked by using data from before 2000 to see if their predictions of disasters materialized during 2000–2015. If a climate model predicted a low probability of a major event during those years and a major disaster did occur, the quality of that model can be contested.[4]

As climate modelers make progress, some of these academics will be content to publish their findings in peer-reviewed academic journals while others will seek to make profits from selling a newsletter. In the financial and real estate sectors, many analysts publish reports predicting the short-term future (for example, "The Dow Jones Industrial Average will rise 20 percent next year") and offering an explanation for their predictions. These analysts compete against one another to develop a reputation for offering high-quality predictions. In other markets such as cars and restaurants, higher-quality products (think of the Mercedes) command a price premium and still attract customers.

The number of firms offering climate prediction services is growing (examples include First Street Foundation, Jupiter, Coastal Risk Consulting [CRC], and Four Twenty Seven) because demand for this information is rising. As an example, CRC sells single-family residence reports for $49 and multifamily and commercial property reports for $499. These climate risk modeling entities exploit spatially refined risk data with powerful computers and employ climate scientists to predict the geographic distribution of new climate risks. In the case of CRC, this entity also offers referral services to connect property owners with consulting and engineering firms to help offset the identified climate risk.

As these climate risk modelers generate location-specific predictions, how will the public judge the quality of their projections? To verify the quality of climate risk prediction, arm's-length climate science experts must be allowed to study the methodology and data sources used by each climate prediction team.

Recent US history suggests that natural disasters have certain clear geographic patterns. In May 2018, the *New York Times* reported that the same areas are hit again and again: "In the last 16 years, parts of Louisiana have been struck by six hurricanes. Areas near San Diego were devastated by three particularly vicious wildfire seasons. And a town in eastern Kentucky has been pummeled by at least nine storms severe enough to warrant federal assistance."[5]

The Federal Emergency Management Agency (FEMA) has created a database spanning 1959 to 2018 that lists which US counties experienced a natural disaster so severe that a disaster was declared.[6] Based on these data, there are clear geographic patterns. During these years, 7 percent of all flood emergency events occurred in Iowa counties, but only 2.3 percent occurred in New York counties. This predictability of what types of natural disasters occur in which geographic areas helps us to cope with such future events.

The Challenge of Predicting Future Climate Risks

A major challenge is predicting the likely timing of when these elevated risks will occur. Many people are worried that parts of Miami face significant sea level rise. Although we can pinpoint the geography of areas at risk for flooding, we do not know if the threat will be really bad for such areas by the year 2030 or by the year 2060 or even later. This timing question is extremely challenging. Climate science predictions about future risks hinge on the world economy's total greenhouse gas emissions growth. This in turns depends on what incentives and rules we have in place. Thus, today's debate about enacting carbon mitigation policies affects the predictions of climate science.

For example, one forecast predicts: "Global average sea levels are expected to continue to rise—by at least several inches in the next 15 years and by 1–4 feet by 2100. A rise of as much as 8 feet by 2100 cannot be ruled out. Sea level rise will be higher than the global average on the East and Gulf Coasts of the United States."[7] These forecasts hinge on an assumption about what the world's carbon dioxide concentration level will be in the year 2100. If we manage to sharply reduce our greenhouse gas emissions, then such a sea level forecast is likely to overstate future effects.

Today, how can we predict the world's carbon dioxide concentration level in the year 2100? The year 2100 is eighty years from now. Think back to a time eighty years ago—the early 1940s. At that time Franklin Roosevelt was the president. Could people back then anticipate the challenges and opportunities that we now face?

To predict the world's carbon dioxide concentration in the distant future with accuracy requires predicting both how many people will live on earth for each year out to the year 2100 and what their average per capita income will be. Such predictions will yield the world's total income each year. If researchers could also predict total greenhouse gas emissions per dollar of income (carbon intensity) per year, then they can predict the flow of annual greenhouse gas emissions produced. The world economy's creation of additional greenhouse gas emissions each year equals average emissions per dollar of output multiplied by total dollars of world output.

Predicting both population and per capita income is challenging. Economists have argued that demographers overstate future world population growth because such demographic models underemphasize the role of human capital, urbanization, and women's labor market opportunities. As people obtain more education and move to cities, they tend to have fewer children.[8] As demonstrated by the deep recession now playing out around the world during the unpredicted covid-19 pandemic, predicting future incomes is a difficult task. Macroeconomists have contentious debates about the expected future economic growth rate in nations ranging from the United States to China to India.

Even if researchers create credible estimates of world population and per capita income dynamics, they must still grapple with the challenge of predicting future carbon intensity (greenhouse gas emissions per dollar of world income). This average carbon intensity hinges on what technologies we are using in the future and what public policies we have enacted. For example, consider predicting what the US carbon dioxide emissions per dollar of GNP will be in the year 2050. The answer will depend on what energy sources we use to generate electricity and to provide basic transportation services. These answers depend on what the price of fossil fuels is versus low-carbon energy at that time. The price of fossil fuels at that time will reflect aggregate supply and demand and whether a given nation has enacted a carbon tax. If the carbon tax per ton equals zero each year, then the economy's carbon intensity will be higher than if the carbon tax is $50 per ton.

The impact of population and per capita income on greenhouse gas emissions depends on the technology used in the economy. Today, the transportation sector creates a large share of greenhouse gas emissions because many cars and trucks and planes run on fossil fuels. In the future, if electric vehicles become the dominant supplier of passenger miles and if these vehicles are fueled using power generated by renewable sources such as wind and solar, then the transportation sector's greenhouse gas emissions will sharply decline.

Even if one could accurately predict greenhouse gas emissions in the future, researchers will still confront the challenge of predicting how this greenhouse gas level will affect a variety of climate measures such as average temperature, the severity of hurricanes, and the distribution of rainfall. This is called the climate sensitivity parameter.[9] This ambiguity over future emissions means that climate science predictions feature great uncertainty about our future.

In climate science today, scientists are grappling with how to incorporate their uncertainty about key parameters into the climate models they create. Climatologist Judith Curry writes: "[Climate models] should be considered not simply as prediction machines, but as scenario generators, sources of insight into complex system behavior, and aids to critical thinking within robust decision frameworks. Such a shift would have implications for how users perceive and use information from climate models and the types of simulations that will have the most value for informing decision making."[10] This is a modest statement. The public looks for precision predictions from our experts. Curry seeks to expand our imaginations about future scenarios that might take place.

Similar issues concerning the precision of scientific prediction have arisen in predicting the death count caused by the novel coronavirus. During the pandemic, the public looked to public health models to predict how many people would die from the virus if there was no enforced social distancing and economic shutdown. As contagion experts used models to generate such an exact prediction (for example, two million lives lost), these uncertain predictions then were taken as fact as policy makers started to make choices (such as mandatory economic shutdowns). This interplay between science and policy would yield better economic and quality of life outcomes if the scientists appreciated how their messages affected policy and if policy makers had some understanding of the scientists' uncertainty about the laws of nature.

The Art and Science of Predicting the Economic Damage Caused by Climate Change

Microeconomic researchers seek to estimate how climate change affects our economy.[11] These academics have been quite creative in collecting data for many indicators of well-being, ranging from agricultural output to suicide rates and to the emotions revealed in tweets and even judges' courtroom decisions.[12] This 2017 paper is one example: "We analyze the impact of outdoor temperature on high-stakes decisions (immigration adjudications) made by professional decision-makers (US immigration judges). . . . A 10F degree increase in case-day temperature reduces decisions favorable to the applicant by 6.55%."[13] Such observational studies are useful for establishing how past climate conditions affected different indicators of well-being. Unlike researchers who conduct a clinical drug trial and choose at random to give some eligible participants the drug while denying it to others (the control group), climate researchers work with observational data. They wait for Mother Nature to launch a natural experiment, such as an extreme heat wave in a temperate place like Berkeley, California, and then they study the effects caused by this event. In such studies, researchers seek to ask "what if" concerning agricultural output, suicide rates, or judges' decisions had the climate event not occurred. To answer this question, researchers study trends in the relevant outcomes during times of typical climate conditions: this represents the baseline for teasing out the damage caused by the climate event. Cause and effect studies are useful for measuring the historical correlations between weather patterns and economic and well-being outcomes.

In my research, I have used a variety of data sets to study how climate conditions affect individuals. Using data from China's equivalent of Twitter, I document that people are in a worse mood (as expressed by the content of their social media posts) on hotter and more polluted days.[14] In another project, I document that the duration of Chinese court cases is longer on more polluted days.[15] My coauthors and I believe that this fact is generated by pollution slowing people's decision making. In other coauthored work, I use cross-country data over the past several decades to document the relation between extreme heat and national economic growth.[16]

Such estimates of the past correlation between climate extremes and economic outcomes intrigue me because they highlight the challenge humanity will face if we fail to adapt in the future. Many young scholars in environmental economics today use these past correlations to predict future outcomes. I have debated many of them about the validity of this approach. I have argued that the extrapolation based on past estimates to the future blindly assumes no adaptation and thus represents a *worst-case* scenario. Such predictions are useful for informing decision makers about the costs they may face in the future if they take no adaptation steps, but they are not likely to be a relevant guide for predicting the actual future that will take place. This is no mere academic debate because it embodies a key idea in modern economics that people, firms, and governments have the capacity to learn and update their behavior.

Let's analyze the standard three-step methodology for predicting the likely damage caused by climate change. Suppose that, using historical data, a research team estimates that the average student's math SAT score is five points lower for every degree that it is over 75 degrees outside during the past month. In step one, the research team uses data to estimate the historical relation between environmental conditions and outcomes that we care about. In step two, the researchers borrow a climate science prediction about future weather changes. Suppose that a specific climate model predicts that it will be four degrees warmer in the future. In step three, the social scientists combine their estimate from step one with the climate science prediction from step two. In the context of the example above, the climate economists would predict that SAT scores will be twenty points lower in the future because of climate change.

This three-step prediction is based on the assumption that the *past rela-tion* between temperature and SAT scores (the slope of −5) will continue to hold in the future. If the world will be four degrees hotter in the future and if each extra degree of warmth reduces test scores by five points, then the change in climate will reduce each student's score by twenty points relative to what each would have scored if the world had not warmed.

This prediction would be more accurate about our future if *no adapta-tion takes place*. But how can this be? If our children suffer from the heat and if our researchers have documented this fact, then in an economy where we are both growing richer over time and air conditioning technology is improving and becoming cheaper, then more people will invest in this adaptation technology. This means that a future researcher, say in the year 2050, will observe that SAT test takers do not score lower when they have been recently exposed to high outdoor heat. This will be evidence of adaptation! The past negative correlation between heat exposure and test scores vanishes.

There is a certain irony that environmental economists implicitly embrace the extrapolation assumption that the future climate damage function will be the same as the one existing today. This extrapolation approach violates the core logic of microeconomics that people respond to incentives! As the rules of the game change, people change their behavior. Environmental economics research actually accelerates this process by expanding people's imagination about the consequences we already face from weather extremes. The popular media actively covers such research. In 2019, my research was profiled in the *New York Post*![17]

The three-stage approach is straightforward to implement, and it simplifies the prediction problem because it abstracts from addressing the thorny issue of how technology changes in quality and price over time. As adaptation-friendly products become more affordable over time, more consumers have the option to buy them, and this helps to attenuate the damage caused by extreme weather. The three-stage approach to predicting climate change's economic impacts does yield fascinating predictions.

One study tests whether extreme heat contributes to the risk of violence.[18] If this hypothesis is correct, then climate change will increase violence and death in some of the world's poorest nations. The researchers collect national data for each nation in Sub-Saharan Africa over the years 1981 to 2002. For each nation, they observe whether a civil war is taking place. They collect additional data on rainfall and average summer temperature. They posit that during hotter summers, the risk of civil war increases. Such extreme weather conditions reduce access to water and crop growth, and rural people begin to fight over increasingly scarce resources.

The researchers document that in the recent past the annual risk of civil war in this region is higher when summers are hotter. In a second step of the project, they use climate projections based on twenty general circulation climate science models to predict the temperature increase that each nation in this region will experience by the year 2030. In the third step of the exercise, they use the historical estimate of the relation between past heat and civil war risk and combine that with the expected increase in temperature. This represents their best guess of the increase in the probability of civil war due to climate change. In the final stage of the exercise, they take the average death count in the past civil wars each year and multiply this by the climate change–induced increase in the risk of civil war. This represents their best guess of the expected death count increase in the future in this region caused by climate change. They predict that climate change will cause an extra 39,455 battlefield deaths per year (in the year 2030) in Sub-Saharan Africa.

This important study highlights how a willingness to extrapolate based on estimates from historical data (the years 1981 to 2002) can be used to make a prediction about the future. An implicit assumption in predicting the future based on historical data is that past correlations will continue to hold in the future. Extrapolation exercises based on past correlations must face the Lucas critique, named after the Nobel Laureate Robert Lucas.[19] The Lucas critique posits that as the rules of the game change, people alter their behavior to do the best they can under the new rules.

The fact that a correlation held in the past can actually help to attenuate this correlation in the future. To appreciate the power of the Lucas critique in the case of this specific example, suppose that there is widespread media coverage of the 2009 paper's prediction of more than 39,000 extra deaths per year on the battlefield due to climate change.

If this scary result is widely publicized, entities such as the United Nations and the Gates Foundation have an incentive to take proactive steps to pursue peace during times of extreme heat. Concerned that climate change will trigger higher levels of violence in Africa, peacekeeping efforts by UN monitors could increase or the Gates Foundation could help to distribute genetically modified crops to help poor farmers grow crops despite increased heat and drought risk. These adaptation strategies (triggered by the understanding of the historical correlation between heat and violence) attenuate this correlation in the future. This example highlights a key theme of this book. It is crucial to document past correlations in order for history not to repeat itself. The research findings (disseminated through blogs and tweets) play a Paul Revere role, sounding the alarm.

Adapting to Anticipated versus Unanticipated Shocks

Consider two different scenarios. In the first, climate scientists predict that we face extreme climate risk and they are correct about this prediction but the public chooses to ignore this warning. In this case, the public could suffer greatly when extreme weather events occur because the public has not prepared for this event. Now consider the opposite case, such that climate scientists predict that abrupt climate change is approaching. Suppose that the public trusts the climate scientists and engages in costly self-protection such as redesigning their homes and buying more powerful air conditioners. In the case in which the public believes the climate scientists but the climate scientists overstate the actual threat, the public will lose money taking precautions that it turns out might not have been worth it (such as more air conditioning in Berkeley). Note the asymmetry that arises. People suffer more if they ignore an accurate worst-case prediction than if they believe a worst-case scenario that turns out to be overstated.

This example highlights how each of us incorporates both advice from experts and our own experiences as we form our beliefs about the future. Such expectations about the future influence our choices today. In Phoenix, Arizona, each summer, the residents expect that it will be extremely hot. The vast majority of people can cope with this anticipated event because they have access to air conditioning and can generally avoid the outdoor summer heat. At a time of rising income inequality, the poor have a reduced capacity to cope with such shocks. Charities and local governments that anticipate this fact can step up their efforts to protect the vulnerable during such times.

When people are surprised by a climate shock, such as a heat wave, then more people will suffer. An example is the Moscow heat wave in the summer of 2010. The people of Moscow were not ready for this event and were caught off guard, and thousands died. After this event, sales of air conditioners soared. If a similar heat wave had repeated the following summer, many fewer people would have died. This is the core test of the adaptation hypothesis. As Mother Nature throws the same punch over and over again, we gain experience and new skills, and access to improved market products, to better handle the shock. The shock becomes less shocking!

The Moscow heat wave case raises important issues of how people update their beliefs once they have been surprised. After the event, how many people expected that such heat waves would occur more often? How many other people viewed this outlier event as simple bad luck that would not occur again? As time passes, does the salience of the shock wear off? If Moscow does not have another heat wave for thirty years, will the people of Moscow in the year 2040 no longer have air conditioning because this costly investment appears to be a waste of money? Or will the memory of the events of summer 2010 linger on and they will remain on guard? I pose these questions to highlight the challenge of learning about how different people form their expectations about the future.

How Do People Form Expectations about the Future?

An ongoing debate in academic economics explores how people form expectations over the likelihood of future events. When I was a graduate student back in the late 1980s, the rational expectations school of thought posited that people use all available information in forming their best guess of the likelihood of future scenarios. Intuitively, this vision modeled people as if they were the logical Mr. Spock from *Star Trek*. In recent years, alternative theories have emerged about belief formation.

The surging field of behavioral economics offers a counternarrative that makes a more pessimistic prediction about how many people update their beliefs about a changing world. In 2017, Richard Thaler won the Nobel Prize in Economics for his work that blends insights from psychology and economics.[20] Behavioral economists argue that people are prone to mistakes. Borrowing from the psychology literature, they argue that many people have traits such as being impatient and thus are unwilling to bear short-term costs for long-term benefits. Some people suffer from cognitive dissonance as they ignore relevant information that challenges their worldview. Others procrastinate and thus are slow to take self-protection steps that will reduce risk exposure. These traits raise the likelihood of being exposed to more risk as such individuals are less likely to see emerging weather patterns. For such procrastinating individuals, environmental science and popular outreach play an extremely useful role by highlighting scary scenarios that we could face if greenhouse gas emissions continue to rise.[21] Imagination facilitates adaptation because it helps us foresee scary future possible scenarios.

The Nobel Laureate Herbert Simon argued that people often follow rules of thumb because thinking and implementing a new solution to a problem are costly. In this case, such individuals will only sluggishly adapt to new challenges.[22] These individuals will tend to rely on what they have seen in the past to form their expectations of the future. For example, a seventy-year-old man might rely on his memories over the past fifty years of how often in summer the temperature exceeds 95 degrees. This would be his best guess of the probability that it will be hotter than 95 degrees next summer. In a world not featuring climate change, this would be a good rule of thumb to follow to predict the future. Based on Simon's theory, such an individual would be slow to update his beliefs about changing climate risks because it is costly for him to process new information. If such individuals believe that the likelihood of future risks will be the same as in the recent past, they are more likely to be caught off guard by changing weather patterns. Such individuals are less likely to take self-precautions or to support elected officials who support greenhouse gas mitigation laws because they do not believe that climate change is a serious issue. If there are many people with incorrect worldviews, then they pose danger both to themselves (by underinvesting in self-protection) and to society (by voting for policies that deliver too low a level of overall public goods targeted to building up resilience).

Research in financial economics has argued that financial investors base their decisions on extrapolative expectations.[23] This model predicts that people believe that the world tomorrow will be similar to the world of the recent past, so when a shocking event occurs, they overreact to the news. In this case, adaptation will feature clumsy stops and starts as people will be too slow in responding to changing conditions and then after a salient shock will panic and over-respond.

Researchers document that people with lower levels of educational attainment are more likely to make systematic mistakes in forming expectations about the future.[24] This finding has two implications. First, it helps policy makers identify which groups of people are at greater risk because they are less likely to anticipate new threats. Second, it suggests that strategies that boost our population's human capital offer an additional benefit that more educated people will be more resilient in the face of future shocks. As nations wrestle with rising income inequality, access to high-quality education is a time-proven strategy for reducing earnings gaps. If more educated people form more accurate expectations of future scenarios, then another payoff of education is greater resilience.

Extra Risk for Climate Skeptics?

In our diverse society, some people are genuinely concerned about the climate change challenge while others appear to not believe that this risk is an imminent threat. For those who objectively face more risk and are increasingly aware that they do face this risk, they will be more willing to invest their money in resilience (by taking actions such as putting their house on stilts), and they will be more likely to support public policies that improve resilience and help to reduce greenhouse gas emissions. The research question here is to understand how different people learn about a moving target. What role do one's friends, government announcements, current events, and news sources each play in helping individuals learn about our changing world?

When one outwardly espouses a belief, this may be due to one's true worldview. A second explanation for adhering to such motivated beliefs is that people seek to conform to the beliefs of their peer group. Consider a new first-year student at Yale University. If this student actively states that he or she does not believe that climate change poses real medium-term risks to our well-being, then this student is likely to be ostracized. In this sense, one's beliefs are a type of badge of honor that determines what groups you join and how you conduct yourself in day-to-day interactions.

In the case of measuring evolving risk perception, an open question is whether so-called climate deniers who state in public that climate risks are small are actively using Google in the privacy of their own home to research the risks they face. In this case, these individuals will be better informed (and better prepared) than is typically acknowledged. Due to motivated beliefs (for example, Republicans who do not want to be seen embracing progressives' talking points), people may make certain statements in public while simultaneously engaging in defensive actions in private that protect their families from new risks.

Whether such media sources as Fox News or the *Wall Street Journal* cause their consumers to tune out and downplay the climate change challenge remains undecided. People can enjoy watching a news source without fully *believing* what they are hearing.[25] News consumers also interact with friends and children and search the web in private. These independent sources of information also influence one's worldview.

Risk Aversion

Our collective choice to allow global greenhouse gas emissions to rise is a type of gamble.[26] When we play roulette or other games of chance, we know the odds. But we do not know the odds of the climate game we are now playing. The probabilities of horrible scenarios increase over time as global greenhouse gas emissions continue to increase.

A risk-averse person is willing to pay to not take a gamble. Economists measure people's risk preferences by observing whether they take gambles (such as investing in stocks versus investing in safer but lower-return bonds). To pin down people's degree of risk aversion, we often ask them to choose among a set of lotteries. For example, suppose I offer you $200 for sure or a risky lottery that offers you a 50 percent chance of winning $400 and a 50 percent chance of winning $0. These two lotteries have the same expected payoff of $200, but the second lottery is riskier. If you choose the risky lottery, then you reveal that you are a risk lover. Now suppose I change the game by changing the first lottery so that you can have $140 for sure. If you still prefer lottery 1 to lottery 2, then I have learned that you are highly risk averse.

Although we know that people differ with respect to their taste for risk, economists have had trouble explaining why this is so. Some research focuses on the role of peers and social preferences.[27] One explanation posits that the historical era of your formative teen years determines your adult preferences. For example, those who grew up during the Great Depression of the 1930s are more risk averse and thus prefer to hold assets that offer low rates of return (say, US Treasury bonds) but are low risk (they pay off at 2 percent per year every year regardless of whether the economy is enjoying a boom or suffering from a bust).[28]

People reveal their risk aversion through their market choices when they hold low-risk portfolios and buy life insurance even when the expected payoff is less than the annual cost. For example, suppose that a woman purchases a one-year life insurance policy for $1,000 that will pay her family $1 million if she dies that year. If, given her age and other demographics, the actuaries predict that she has a one in ten thousand chance of dying, then the expected value of this policy equals .0001 × 1 million = $100, but she paid $1,000 for it. She reveals that she is quite risk averse when we observe her buy this policy.

In my research, I document that richer people are willing to pay more to avoid a risk such as working at a job with a higher chance of fatality such as mining.[29] This implies that poorer people will often take on more risk such as living in homes that face more risk from crime or pollution. This raises deep issues of environmental justice that I discuss in chapter 3. A fair way to reduce such disparities in risk exposure between the rich and the poor is to build more housing in relatively safer places. As I discuss in chapter 10, the US land zoning code often places regulatory limits on how much housing can be built.

A Road Map

As individuals and as a society, we could more easily adapt to climate change if the experts would converge on understanding how the climate system works and then shared their new knowledge with the concerned (and trusting) public. The challenge we face is that the scientists, though making progress, are still uncertain about emerging risks. At the same time some of the public are following the science while others are downplaying the climate change challenge. Environmentalists are quick to accuse such climate skeptics of being climate deniers. An alternative hypothesis is that some of the so-called climate skeptics voice these opinions in public due to their own concept of identity and fitting in with their peers and neighbors. In this case, such individuals will consider taking private actions to protect themselves against new risks even if they do not attribute the cause of these risks to be climate change. Note that this possibility highlights a dichotomy between how people act in the public arena versus how they conduct themselves in private markets. I would not call these people hypocrites. Instead, I view this group to be believers in the power of markets to offset new risks. The set of willing adapters will increase as the price of market goods that promote resilience (think of air conditioners as an example) becomes cheaper over time.

This book focuses on how capitalism helps us to adapt to the climate change challenge through facilitating behavioral change. In the next four chapters we'll explore how households, local governments, and firms respond to emerging climate risks. The job of an economist here is to study how markets for goods, capital, and land adjust in response to demand and supply shifts. Markets expand the menu of opportunities available to buyers and sellers as they pursue their goals.

2

Daily Quality of Life

In the midst of the covid-19 pandemic, many people face both health contagion risk and economic risk as millions are unemployed and small businesses have few customers. The sudden shock caused by the combination of the pandemic risk and the economic shutdown highlight how rapidly comfortable circumstances can change. In this chapter we'll explore several pathways through which people can shield themselves and their families from risk.

Natural Disaster Risk

Many thousands of people have died in natural disasters in recent decades. Climate scientists generally believe that climate change contributes to hurricane, flood, and fire risk. The death toll from natural disasters declines as people and nations grow richer.[1] Richer people have the resources to pay for both public goods and to invest in private goods that better protect them from risks. Richer people live in higher-quality housing that can better withstand floods and storms. In the tornado zones, fewer people's roofs are being torn off by storms. As people grow richer, they also seek to avoid risk. In my joint research, I document that across the decades from 1940 to 1990 in the United States, workers in risky industries such as mining and construction were paid an increasing wage premium over time for risking their lives.[2] At any point in time, richer people live in low-crime areas and buy new, less risky products. Over time, more Americans have grown richer because rising educational achievement builds human capital. As people acquire more human capital through schooling, they become better problem solvers and shrewder analysts of news and information. As the public demands more sophisticated news reporting, suppliers of information such as Bloomberg, the *New York Times,* and the *Wall Street Journal* will invest more resources in delivering more sophisticated news. This refined information helps educated individuals to stay current on emerging challenges and to better anticipate new risks that may affect their family's well-being, assets, and livelihood. In this sense, the rise of real-time 24/7 media facilitates adaptation.

After a disaster, important issues arise concerning the long-term scars to a family's finances and to the mental health of those affected. Researchers have investigated how homeowners with a mortgage respond after a disaster strikes.[3] For such households with insurance and who believe that the area is in decline, they find that homeowners are more likely to pay off their remaining mortgage balance and move away. Others hold the option to default on their loan in the future. (In chapter 7, I discuss how real estate owners respond to natural disasters both before and after they occur.)

After a disaster, as community members seek to rebuild, the rise of modular construction technologies will reduce construction time. Modular technologies open up the possibility of rebuilding homes in six months instead of years. In the aftermath of such a disaster, if landowners now view the risk of future disasters to be higher, they may not wish to build a long-lasting, expensive structure.[4]

Migration

As we better understand the geography of new climate risks, some people may choose to migrate away from such hazards. Each of us chooses where to live and whether to migrate to a new location. In making this important choice we weigh the benefits and costs of each possible destination. Urban economists point out that people choose a location (say, Miami or Houston) based on its labor market prospects, real estate prices, and quality of life and taxes. A city that features a great job market and great quality of life and low taxes will have to feature high real estate prices. Why? If real estate prices were low there, everyone would move to this great city and this would bid up real estate prices. Which areas have great quality of life depends on one's tastes. A sunbather will rank Miami above Minneapolis, while a cross-country skier will have the opposite ranking. Americans have revealed a taste for warm-winter, less polluted, low-risk areas, and these locations feature more expensive real estate (discussed at length in chapter 6). Once people have selected a city to live in, they also choose a home and a neighborhood. In daily life, people consider both the price of housing in a given location and the location's implications for the commute time to work, commute time to the consumer city, local public school quality, and local crime levels and pollution levels.[5]

Rising climate change risk will shake up the rankings of America's best cities as some areas will face greater challenges than others. As people begin to move to the safer locations, this rising demand could lead to sharply higher real estate prices in these locations. This raises the concern about environmental gentrification—real estate price dynamics brought about by shifts in demand. Economic logic suggests that the extent of such a migration-induced effect hinges on supply. As demand to live in a given set of relatively safer locations increases, can real estate developers build more housing there? Or do land use zoning policies inhibit such new construction? Stringent land use zoning in safer areas will translate into higher real estate prices in those areas and will lower the likelihood that poor people can migrate there. I return to each of these points below.

Consider that the supply of safe locations and housing structures is determined not only by natural geography (the topography and elevation of land) but by private sector developers (how structures are built) and by government policy (zoning codes and engineering projects). The key issue to note is that governmental rules are not set in stone. Those who use historical data to judge future effects of climate change are implicitly assuming that government regulations do not become more increasingly favorable of resilience over time.

Demographers and economists have noted the rise in positive assortment of adult couples such that highly educated people tend to live with a highly educated partner and less educated people pair off with a less educated partner. In 2000, I coauthored a paper on the locational choice of highly educated people as we studied the propensity of so-called power couples to choose to live in big cities.[6] Such cities offer a larger local labor market that presents more work opportunities for each partner to find highly specialized jobs. The net effect is that neither sacrifices economic opportunity in the name of remaining a couple when they move. When each educated member of the couple can find a good job in the same location, the household's total income will be even higher. Richer people can afford higher-quality products, and this will improve their quality of life and shield them from risk. This case highlights how cross-family variation in exposure to risk increases as a function of the choices we make. If partners were randomly assigned to live together, there would be greater climate justice because there would be fewer poor adults. A household's income plays a central role in protecting it from climate risk as richer people can live in safer areas and in safer structures and have the financial freedom to upgrade their home to reduce its risk exposure.

Migration Costs

Those who face high migration costs will be less likely to leave their current location. Such costs will hinder adaptation for those who live in areas facing greater natural disaster and heat risks. People who cannot move are tied to that location. In this sense, the shocks that affect the place also affect these people. Those who can easily move always have the option to leave if the location is at increasing risk to suffer from a severe shock or has recently experienced a severe shock.

Imagine a Houston family that rents an apartment. The family has good friends in Dallas, and the adults can easily transfer to work in Dallas. This family suffers little from shocks to Houston (such as Hurricane Harvey) because it can simply move to Dallas. Because the family is renting the apartment, the household does not suffer an asset value loss when the shock damages the property. In this case, the family's labor market skills and human capital, its social networks and asset portfolio, allow it to cope easily with the shock that hits Houston. The ability to get up and go offers this family flexibility that becomes even more valued as climate risks increase.

To flip this example around, consider a family that chooses to plant roots and commit to remain for decades in the same city and the same community within that city. This family is making a long-run bet that this community will continue to thrive. Although such a place-based bet offers the family psychic benefits, it also exposes the family to more long-run risk because this family is sacrificing the option to move at a time when climate change is increasing place-based risk.

Since migration is costly, young people are more likely to move than older people because they have a longer-term horizon to gain from having more time spent working and learning in the new destination they choose. Migrants are moving themselves and their families to new local labor markets with different qualities of life. Young people have strong incentives to compare different possible locations based on their employment and earnings opportunities in each potential destination and what their quality of life will be like there. Even among young people, they will vary with respect to their ability to anticipate future risks that they may be exposed to in a location where they have never lived. Recent research has emphasized the importance of cognitive skill and imagination in anticipating likely risk exposure.[7] This research makes a pessimistic prediction that people with less skill and less imagination will be more likely to underestimate future place-based shocks and thus expose themselves to more risk. In this case, climate change will increase inequality.

Economists think about moving costs in a distinctive way. We acknowledge that migrants must pack up and pay for the move, but we also factor in the intangibles of what they lose by leaving their current location. Suppose that Susan has lived in the same Houston neighborhood for thirteen years. She is revealing that she values living there. Suppose further that Susan's current neighborhood has several attributes that she appreciates. Examples might include that she has a short commute to work, her church is nearby, and she likes the local restaurants. If Susan seeks out similar attributes in possible destinations she is considering moving to, she will have a hard time finding similar good matches. This is known as the curse of dimensionality. Other places to live may feature good nearby restaurants but her commute to work would be longer and she would have to travel far to attend church. In this case, she would have trouble finding a near match for her original neighborhood and she will be less likely to leave it even if natural disaster risk or other climate disamenities make it less pleasant. In Houston, Susan has found her favorite neighborhood niche. Such a well-settled person will be less likely to move even if climate risk increases.

How did Susan's good match with her current location occur? Did she always have a taste for the locational features that her Houston neighborhood offers? Or as she lived in this community, did she gain an appreciation for its ways and develop location-specific social capital such that she has now locked in and now has a strong preference to remain there? For those who have built up location-specific capital, they know that they will lose this if they move to a new location. This discourages such outmigration and places them at greater risk of injury by place-based shocks such as natural disasters.

In the language of economics, this Houston resident is not close to the margin. Consider the following example. If you love Starbucks coffee and would pay up to $12 for a cup, then if Starbucks raises the price from $2 to $2.40, you will continue to drink Starbucks coffee. In this case, you are not at the margin. If, however, Starbucks raised the price from $2 to $13, you would no longer purchase Starbucks coffee.

Place-based attachments reduce the likelihood that people are willing to move. Residents of New Orleans who love the city's history and culture will be less likely to leave even if New Orleans experiences bad shocks. Evidence supporting this claim is provided by the natural experiment created by Hurricane Katrina in 2005. Some families left New Orleans because of the wreckage caused by this shock. Many of these displaced people enjoyed an income gain (as revealed by subsequent Internal Revenue Service tax return data) when they moved to cities such as Houston. This research suggests that people were implicitly paying (in terms of the extra income they would earn if they moved to Houston) by remaining in New Orleans. Economic logic says that both their social ties and their love of local New Orleans culture kept them in a location that did not fully reward them for their skills.[8]

Coping with Change over the Life Cycle

Middle-aged and older people will face greater challenges adapting to new climate risks than young people. The middle-aged have already locked in to an area as they carefully made a tradeoff decision balancing the location's proximity to work, schools for their children, access to their parents and friends, and leisure opportunities. If they must now move, they will have trouble finding another city offering similar attributes along all of these dimensions.

Consider the example of Frank, who lives in Seattle. He is fifty years old and has lived in Seattle for twenty-five years. Frank works at Amazon and has friends who live in the area. His leisure and sports hobbies are tied to Seattle's seasons. He has planted roots there and has local friends and a connection to the place. His next favorite city to live in (perhaps San Francisco or Boston) is not a close substitute for his Seattle. It will be easier to adapt to climate change if everyone has a second choice that is a close substitute for their current location. If future Seattle no longer has a cool summer, then Frank will have to change his leisure routines. Seattle will still be Seattle, but climate change will have shifted its attributes and he will not easily be able to access "his old Seattle."

Consider this statement about Seattle by *New York Times* columnist Timothy Egan: "As a native Seattleite, I've always wondered what it would be like to live in a place where it's sunny every day. Now that I'm experiencing something close to that, I feel out of sorts in a strange land. Wildfires burn today in the Olympic Mountains west of Seattle, a forest zone that is typically one of the rainiest places on earth."[9] To an economist, the key question here is: What has he lost due to climate change? What would Egan be willing to pay to have his old Seattle back? From his perspective, there are no close substitutes for his Seattle in the rest of the United States. He is expressing sentimentality for his past.

A family's demographic structure will also play a role in determining its ability to migrate. Younger family members often provide care for their aging parents, and grandparents often provide child care for their grandchildren. These intergenerational interactions tie a family to a given location. The young mother may rely on her mother for child care for the grandchild. Her alternative is to go to the market and seek child care from strangers. If the grandmother does not want to move away from the area because her friends and roots are there and the young mother is eager to have her mother provide child care, then the young mother is also tied to the local area.

Ongoing research studies the mental acuity of senior citizens.[10] The elderly make key decisions concerning how to invest their savings and how they handle their health care and housing assets. Some suffer from memory loss and some are diagnosed with Alzheimer's. Those suffering from these conditions will have a reduced capacity to cope with new risks. This imposes extra costs on their children (often their daughters) to help them through day-to-day activities.[11]

This example leads to the issue of transportation speed within cities. In Singapore, for example, people can always move at forty miles per hour because Singapore has enacted road congestion pricing. During high-demand times of day such as rush hour, the congestion charge increases, and this reduces road use as some people substitute trip modes and use public transit instead. The net effect is that road speeds do not decline at peak times. If a young woman's mother is able to move at forty miles per hour and she is willing to commute thirty minutes each one-way commute to provide child care for her grandchild, then this means that the young mother and her mother can live twenty miles apart and still make this work. The ability to move at higher speed within a city increases the menu of possible geographic locations from which a family may choose.

Friendship with people who live in the city where you live and work ties you to an area. Such ties may help you to adapt in the middle of a natural disaster because your friends are looking out for you and working as a team to survive. Alternatively, local friends may act to tie you to a specific area, and this increases your risk exposure. When people make friends with their neighbors, this makes them more likely to stay in such a community because if they move they will incur new costs to make new friends. This local tie to one's friends can reduce one's ability to cope with place-based risks because such a person with many local friends will be less likely to move away to an objectively safer place.[12] Of course, the win-win solution is to remain near your social network while finding a place to live that is safe (that is, on higher ground).

For those who are tied to a geographic place because of a love of its culture and amenities or because one's family and friends continue to live there, recent technological innovations reduce the labor market penalty for not living in the best local labor market. Back in the year 2000, workers tied to New Orleans who were offered great jobs in Boston would be likely to turn down the Boston jobs because of their loyalty to New Orleans. Such individuals would sacrifice the extra income they would have earned if moved to Boston. Given that higher incomes offer greater quality of life in part because of the extra safety and resilience that can be purchased with extra resources, these people were sacrificing. In the year 2020, telecommuting and the rise of web-based conferencing technology, such as Zoom, created the exciting possibility, especially for educated workers, that work can increasingly be done from home. This means that the tied New Orleans workers can have the best of both worlds by living there while working for Boston firms. In this case, the information technology unbundles place of residence from place of work.[13] It is important to note that research indicates that more educated people are more likely to work in jobs that can be done at home or remotely. This issue has arisen in the context of the ongoing covid-19 pandemic as people such as college professors can work from home and Zoom to deliver course lectures. In contrast, less educated workers, such as those who work at grocery stores, must go to work to do their job and thus face greater contagion risk. These facts highlight how, during times of risk, existing inequality gaps actually widen.

Even as we recognize that technological innovation does not equally benefit all people, we should note that the ability to telecommute opens up many exciting adaptation possibilities because it sharply increases the different permutations of choices that people can consider. For example, a highly risk averse person who can telecommute can move to a very safe place while telecommuting to a place that may be located in a riskier setting.

Mitigating the Health Impacts of Climate Change

Our exposure to pollution and heat depends on where we live, how much time we spend outside, and the investments we take to reduce indoor heat and pollution.

Even when outdoor pollution levels are high, people can reduce their exposure to such pollution by purchasing air filters and masks. China's cities are often highly polluted. To protect themselves from such pollution, many Chinese urbanites purchase products on an Internet platform called Taobao (the Chinese equivalent of Amazon). My coauthors and I studied data we collected from Taobao for thirty-five Chinese cities during the year 2014 on daily sales of air masks and air filters. Masks are cheap. Air filters are more effective and more expensive. We were able to segment consumers into three income categories (poor, middle class, and rich). We documented that each income group buys more masks on more polluted days, but the rich are more likely to buy air filters on more polluted days. Although we showed that Chinese urbanites spend more on masks and air filters on more polluted days, we do not know how effective these products are in offsetting pollution and thus how much sickness is prevented.[14] Future research could study this to investigate whether increases in pollution in China's cities are causing less sickness over time (for both the rich and the poor) than they did in the past. If the correlation between outdoor air pollution and population sickness declines over time, then this will be direct evidence of an increased ability to adapt to the pollution threat.

As climate change increases average temperatures, such increases will lead to higher levels of ambient smog pollution.[15] To offset such increased risk, people will need to spend less time outside on hot, polluted days. People with indoor jobs will have a greater ability to adapt than people who work outside. More educated people are more likely to have these jobs, and this suggests that climate change will increase quality of life inequality.

Higher temperatures are also associated with negative emotions. A creative study documented this fact using social media content expressed in billions of Twitter tweets.[16] Although every tweet on social media differs, linguists have figured out an algorithm that uses the information coded in the words people use to assign a given tweet a "sentiment score." Such a score indicates whether one is in a good or bad mood. Based on this metric, research documents that on hotter and more polluted days, the population is in a worse mood. In my recent research, we document similar patterns using social media data from China.[17] Another study shows that rising temperatures in the United States and Mexico are associated with higher suicide rates in both nations. The US results were based on collecting county-level data such that the researchers observed each county's suicide rate in different years and correlated the dynamics of this variable with changes in the county's temperature.[18]

Each of these three correlation studies documents the past economic damage caused by extreme weather conditions. Such new knowledge can help to protect people from suffering similar damage in the future. The key social science question is whether people are aware of how such extreme heat affects them. If those who are susceptible to extreme heat have a better understanding of these effects, then it is in their self-interest to change their behavior to reduce their exposure to such conditions. We are not passive victims here! Although such behavioral change could impose some costs, this research highlights that there would be benefits. These correlation studies fuel adaptation behavioral change, and this means that future researchers will find small effects of future heat on Twitter sentiment. Why? The Twitter study documents that on extremely hot days, people were in worse moods. If people read about this study's results (on their Twitter feed perhaps!), and if this leads them to take steps to relax on hot days (because they recognize that they are susceptible), then in the future when it is hot, they will be less angry and the future research study will document a smaller correlation between outdoor heat and anger. This would be direct evidence of adaptation. Note that it is not the case that people's biology is changing such that they can withstand the heat. Instead, the true dynamic here is just the opposite. Since people will have learned from the research that they cannot take the heat, they will take proactive steps to protect themselves, which will allow them to avoid the challenge that they now anticipate.

An extension of this important research would be to collect individual-level data on suicide propensities rather than tracking geographic aggregates such as counties over time. By accessing such individual-level data, researchers could identify which subset of people are on edge at a given point in time (perhaps because they have lost a job or broken up with a partner) and can be pushed over the edge by high levels of heat and pollution. This example highlights the nuanced hypotheses that can be tested because of the big data revolution. In our diverse society, certain types of people may be at extreme risk on hot days while other demographic groups may be at little risk. In this sense, there is no average person. One of my teachers at the University of Chicago often said: "If your head is in the oven and your feet are in the fridge, then on average you are 'okay.' " He sought to have us focus on the distribution effects of any given cause rather than on an often fictitious average person.

This point is crucial: the heart of the adaptation challenge from the perspective of microeconomics is to understand how different people cope with the same challenges. Let us return to the research that documents the association between high outdoor heat levels and increased risk of suicide. Given how few people commit suicide each year, it is likely that only a small group of people are at risk at any point in time. A just society would focus on identifying those at greatest risk and targeting mental health resources for them. Such efforts would help these individuals to adapt to the new climate risks. Facebook is profiling those at risk for suicide based on the language that people use on the Facebook platform.[19] A statistical research project that correlates this Facebook indicator with outdoor air pollution and outdoor temperature could quantify how the mood of the same person changes over time as weather patterns change. These insights could then be used by public health professionals to step in and introduce interventions to protect such vulnerable people. Such proactive efforts would reduce the correlation between suicide risk and exposure to extremely hot temperatures.

This example highlights how research on the past correlation between suicide and heat attenuates this correlation in the future if we learn from this research and target resources to those who are at the greatest risk. If higher temperatures increase suicide rates only slightly and if this correlation is unknown by public health officials, then it is unlikely that society will respond to this emerging challenge because we are unaware of this challenge. In contrast, if credible research documents a robust association between extreme heat and violence, then this research creates the possibility that public health officials will take proactive steps to protect the vulnerable on such days by contacting them and by nudging the family and friends of such vulnerable individuals to check in with the specific person. In this sense, higher temperatures have a direct effect on raising violence, but the expectations of such higher temperatures increases the likelihood that the state takes preventative actions. If access to mental health care increases during heat waves, then the correlation between higher temperatures and violence will be lessened by the increase in resilience investment. The understanding that there has been a past correlation between heat and suicide leads to more proactive outreach by friends and public health workers, and together these efforts mitigate the heat and suicide correlation.

We are not passive victims in the face of climate change. If we understand how we have been injured by extreme weather in the past, we can devise new low-cost strategies to reduce our risk exposure going forward. This adaptive response to an anticipated threat reduces the social cost of climate change.

Staying Healthy in the Face of High Heat

In the past, before the widespread adoption of air conditioning, high heat exposure raised mortality rates. Using data from both India and the United States, researchers document that the correlation between mortality rates and high heat is declining over time.[20] Public health research on long-run trends in New York City shows the increased capacity of this city's residents to cope with extreme heat.[21] This means that fewer people die on extremely hot days now than in the past. The main reason for this trend is the increased access to air conditioning. The expanded use of air conditioning over time is mainly driven by rising incomes, falling equipment prices, and a greater awareness of the challenge of heat exposure.

Although an optimist can point to these positive adaptation trends, many urbanites continue to lack access to air conditioning. An examination of fifteen major US cities concludes that future warming could cause increased heat-related mortality risk if adaptation steps are not taken.[22]

In temperate Berkeley, California, today, the average August high temperature is 72 degrees Fahrenheit. Most homes in Berkeley built before 1950 do not have air conditioning; they were built before air conditioning became widespread. Given the area's cool summers, few residents are likely to view installing air conditioning as a worthwhile investment. As summer temperatures rise, this group is taking a gamble if they choose to not install air conditioning. My wife's mother lives in a Berkeley home that does not have air conditioning.

Sacramento is close to Berkeley, and in this warmer region most homes do have air conditioning. The people of Sacramento do not have different preferences than the people of Berkeley. However, they do have different expectations concerning how hot local summers will be. Returning to the expectations theme of chapter 1, the people of Sacramento expect that the summers will be hot, and they have taken proactive steps to prepare.

If Berkeley's daily temperature shifts so that hot days (say, over 85 degrees) become increasingly likely and the area more closely resembles Sacramento's current temperature distribution, then my mother-in-law would be more likely to install air conditioning. In this case, the expectation that the Berkeley climate is changing will cause her to invest in a product that shields her from climate risk. This dynamic means that she will mainly face risk in the short run during the time when the climate is warming. However, she has not retrofitted her house for air conditioning because of the inconvenience. If she sold her home to a young family, the new family would be highly likely to air condition the home because the retrofit is a one-time fixed-cost investment that will offer benefits to that family for the next decades. This example highlights that transactions in the housing market (as older people sell a home to a younger family) actually increase the real estate stock's resilience because new buyers are more likely to upgrade the resilience infrastructure. Markets promote resilience.

It is interesting to contrast the retrofitting decision regarding Berkeley's older homes with air conditioning installation rates in China as this nation's urbanization continues. Many climate change models predict that parts of China will see a large increase in extremely hot days. In China, newly built housing units are equipped with up-to-date air conditioners that protect residents against weather extremes. Today, urban Chinese households own an average of 1.2 air conditioners. In richer cities, this ownership rate is even higher. More buildings will have central air conditioning, and this will protect the population from heat waves.[23]

A third case study is Karachi, Pakistan.[24] It is extremely hot there, and air conditioning ownership rates are low. To run an air conditioner requires electricity, and the Karachi power grid is unstable and unreliable. When a key private good such as air conditioning requires as an input electricity and the electricity is provided by the public sector, the public may not be able to protect themselves. The ability to engage in private self-protection hinges on the quality of the public infrastructure power grid. There is a danger of granting a public monopoly to a single entity because this lack of competition weakens the incentive to deliver a reliable product. As decentralized power generation becomes more feasible through wind and solar generation and cheaper backup power generators, these power reliability issues may fade, but this is a crucial issue for facilitating household adaptation.

Feedback Loops

A concern about increased air conditioning in our hotter world is that this creates a feedback loop such that greenhouse emissions will rise as more power is consumed to power our air conditioners. This dynamic would exacerbate the climate change mitigation challenge. One study documents that Mexico's total electricity consumption from the residential sector will soar as the growing Mexican middle class catches up to the US middle class in appliance ownership rates.[25] But it is important to note that the residential sector represents roughly 15 percent of total electricity consumption. A large increase in electricity consumption for a relatively small sector of the economy has only moderate implications for exacerbating the carbon mitigation challenge.

To measure the private costs of adapting to increased heat through air conditioning, a researcher would need to know the annual cost of owning this durable and the electricity expenditure to operate this device. To measure the social costs associated with this privately beneficial activity, the researcher would need to estimate the emissions associated with this extra electricity consumption. If the power is generated using renewables, then this social cost would be low. If the power is generated by coal-fired power plants, then the social costs would be higher. Economists have studied the social costs of driving electric vehicles.[26] In areas such as Ohio, where the power is generated by coal, driving a conventional gasoline vehicle can be cleaner than driving an electric vehicle such as a Tesla.

A growing share of overall power generation is coming from such renewable sources as wind and solar. The rise of global renewable power generation raises the possibility of achieving the win-win of adapting to climate risk while also slowing our new emissions of greenhouse gas emissions. The phase-out of coal (and the rise of renewables and natural gas as electricity power sources) has meant that the negative externality from driving electric vehicles and using air conditioning is declining over time. In the world's largest developing nations, China and India, increases in electricity consumption and in sources of power (for example, coal versus renewables) play a key role in determining global greenhouse gas dynamics. Until recently, more than 70 percent of the power generated in both nations was produced by coal-fired plants. As global technological progress in renewable power takes place and more middle-class people in China and India are affected by the air pollution generated by coal-fired power plants, these nations are more likely to substitute away from coal-fired power plants.[27] As this dynamic plays out, the social cost of greater reliance on air conditioning declines.

Adapting to Extreme Climate Conditions

Research documents that when people live in areas with a history of being exposed to extreme heat (think of Phoenix or Las Vegas), the households who live in these places and their local governments have made investments to be ready for the heat. In such cases, write the researchers, temperature extremes cause much less suffering and mortality is reduced: "Using exogenous variation in temperature and data on all elderly Medicare beneficiaries from 1992–2011, we show that the mortality effect of hot days is much larger in cool ZIP codes than in warm ones and that the opposite is true for cold days. We attribute this heterogeneity to historical climate adaptation. As one adaptive mechanism, air conditioning penetration explains nearly all of the regional heterogeneity in heat-driven morality but not cold-driven mortality."[28]

Phoenix-area real estate and homes are ready for extreme summer heat but not for extreme cold. The opposite is true in Vermont. Real estate owners in Vermont can learn from the Phoenix experience to see what might be cost-effective strategies to deal with future expected heat. Historically cold places that expect to be warmer in the future can learn about adaptation strategies today by researching how hot cities have planned their capital stock. If a place such as Vermont now faces both extreme heat and extreme cold, then new architectural challenges will arise in building structures that can handle both extremes.

A supply-side solution to this challenge is for national home builders to build more homes. On its webpage, the nationwide builder KB Home touts the energy efficiency features of its homes.[29] A national home builder builds in a variety of climates. This cumulative learning by doing helps it to build housing that protects families from the specific location conditions they face. This example highlights that human capital and learning by doing facilitate adaptation. An architectural challenge will arise for areas that will suffer from both more extreme heat and extreme cold due to climate change. Architects and engineers will have to wrestle with whether homes that are prepared for extreme heat are at a disadvantage (in terms of higher cost) to handle extreme cold as well.

Extreme Air Pollution Spikes

Climate change poses increased drought risk in the US West. Over the past few decades millions of people have moved to California, Utah, Nevada, and other western states, increasing demand for scarce water. The combination of heat and drought is also increasing the risk of significant wildfires in the West. As more people live in fire zones, they increase their risk of exposure to air pollution created by wildfires. The fires in Southern California in December 2017 and November 2018 made the national news. Residents of Los Angeles faced high levels of air pollution caused by the fire that started on December 4, 2017. Using data from the Environmental Protection Agency from Santa Barbara County for one air quality monitoring station, I find that starting on December 5, local air pollution levels soared and that they returned to normal only ten days after the fire started.

On such smoky days, day-to-day life is unpleasant. Staying inside partially insulates people, but the smoke eventually permeates homes and offices. During those days in late 2017, I was in Los Angeles and I experienced headaches. To measure the social costs of such unpleasant events, one could measure the extra mortality risk and morbidity risk experienced on those days, but such health costs would underestimate the social losses because people such as myself who exhibit no symptoms also suffer on such days. Another method for measuring the costs of air pollution is simply to survey a representative sample of people affected by a fire to learn how much each of them is willing to pay to not be exposed to pollution. Since millions of people in Southern California are all exposed to these losses, the total social cost can be high.

The rising challenge of western fires creates new opportunities for firms that can innovate to provide products that improve quality of life in the face of these environmental conditions. Consider home air filters: the Alen BreatheSmart FIT50 air purifier is at home in the family room and offers nine hundred square feet of coverage on its highest speed.[30] It costs $549 and claims to remove 99 percent of particles less than 0.3 microns.[31] If this and similar products are as effective as advertised, then they can help to reduce the damage caused by wildfires. Returning to the concept of the economic damage function, empirical researchers can test this claim.

Suppose such a research team could randomly distribute these air filters to some households and not to others. The former represent the treatment group while the latter represent the control group. The research team could track key outcome indicators such as adult worker productivity and teenager test scores and compare these outcomes during the wildfire season for those families in the treatment group and the control group. If these products are as effective as advertised, then the families in the treatment group will not experience reductions in productivity while those in the control group (who are less prepared for the wildfire smoke) will suffer more. The average performance gap (calculated based on the average productivity of those in the treatment group minus the average productivity of those in the control group) would represent a key measure of the benefits of investing in such self-protection. After the study's results are measured, this information can be shared with the public. Even without government subsidies, people will purchase such equipment if the private benefits to them exceed the costs. This example highlights how research fuels adaptation. When a market product can credibly be demonstrated to be effective, more people will buy it, and as more of the population owns such a product (in this case, the BreatheSmart FIT50), the negative correlation between productivity and outdoor air pollution will diminish. This example highlights the productive role that markets play in insulating us from risk.

Leisure and Climate Change

Most Americans have several hours of leisure on both weekdays and weekends. How will climate change affect how we use our leisure time? Let's consider the case of Minneapolis, where climate change will raise local winter temperatures. Warmer winters may reduce mortality risk but could reduce outdoor winter leisure opportunities for groups such as cross-country skiers.[32] For these skiers, quality of life in Minneapolis will decline as climate change melts the snow. The key microeconomic question here concerns what these people will substitute for skiing and how close a substitute for skiing the new activities will be. In a future world where Minneapolis features less snow, will the next generation of young people find new leisure activities? In the case of leisure, the cost of climate change revolves around how well these new hobbies substitute for cross-country skiing. Because the young have less experience skiing, they will face lower adaptation costs. For the middle-aged, who will be old in a few decades, there is an open question of whether they still can ski and what they will substitute. If their next best alternative is watching television and they dislike television, then they will suffer more than if their next best alternative is doing yoga. A fifty-year-old man who loves cross-country skiing may discover a new passion for yoga. If the main joy from the leisure activity is working out with friends, then the ability to convince his friends to join him in the yoga class will lower the cost of switching from skiing to yoga. If the fifty-year-old's pleasure from skiing is really based on working up a sweat and strengthening his muscles while being with his friends, then he can find close substitutes inside. If this man really loves only cross-country skiing and nothing else gives him pleasure, then he suffers more if can no longer ski. The key point here is to consider what we lose when our past favorite choice is no longer available. A flexible person will discover, and through investments of time and effort develop, other leisure hobbies that could be close substitutes for one's original leisure activity.

Climate change will alter the attributes of geographic locations. For the people who are long-term residents of these places, they will suffer more from local changes because they were so well matched to a place that no longer exists. Consider the case of there being less skating on ice in Canada because the ice does not stay frozen in a future warmer winter. As one man told the *New York Times:* "There's a huge difference between when I grew up and was skating outside, and the last five years of skating out here. . . . Will my kids, my grandkids, be able to play in an outdoor rink? Probably not. It might be a dying tradition."[33] This man hopes that his grandkids will have access to similar future activities that he enjoyed in his past. He is upset that his old ways are becoming extinct. But it would be an exaggeration to leap from this sentimental point to the claim that his grandkids will have nothing to do in a warmer winter in Canada. Way back before Google and the Internet existed, I did not anticipate the role that YouTube would play in my son's life.

If Canadian parents want their children to develop their social skills and to exercise and if skating is now not available, then the issue arises whether indoor basketball or something else is a close substitute. In a world with many alternatives, the next best alternative must be a close substitute in terms of achieving core goals of achieving health and comfort. If most people easily substitute to another leisure activity, then they lose little from the changing conditions.

At any point in time, three age groups are alive: the young, the middle-aged, and the old. The middle-aged and old will remember the old Minneapolis, and they will lose as their Minneapolis vanishes. But the young never lived in the old Minneapolis and thus cannot miss it.

This ability to substitute leisure activities may differ across generations within the same family. Consider a father and son. The father likes to jog, and he loses the ability to engage in this activity on an extremely hot day. His son is a gamer who loves video games. On a hot day, the son stays at home and watches a gaming competition. The father loses from the heat, but the son loses much less.

The extra costs imposed by climate change in terms of lost leisure opportunities differ across people. Consider a chess player who plays chess inside. The chess player's leisure is insulated from extreme heat. In contrast, a jogger will be exposed to high heat and will suffer greater discomfort. Our enjoyment of our time depends on outdoor conditions.

When such outdoor weather conditions are predictable, people will rework their schedules to adapt to the local conditions. In Singapore, a city with extreme heat and humidity, the city comes to life late at night when the temperature has dropped. In European nations with hot summers, the siesta allows people to avoid the hottest part of the day. We can rework our schedule to accommodate anticipated extreme events as a way to reduce the costs of extreme conditions. Climate forecasts now provide us with an accurate seven-day forecast of the weather we will face in a given location. Such forecasts help us to make plans to spend time outside when it is more pleasant. Tennis players who enjoy playing outside will have more trouble booking a court at such times because many others will have similar plans. If there is a market for allocating such courts, then those who value them the most will be able to play (but they will pay to do so).

In my daily routine, I reallocate my outdoor time to cope with the heat. The Beverly Hills Farmers Market is open on Sundays from 8:00 a.m. to 2:00 p.m. On extremely hot days, shoppers arrive at 8:00 a.m. when it is cooler outside. On such days, rich people in Beverly Hills will postpone golf outings for a cooler day or a later tee-off time. This suggests that people will have many choices over how to reconfigure their leisure and shopping to reduce their climate exposure. In a future world of automated vehicles and drones, one can imagine a near-term future where people spend more time inside as machines bring them goods, thereby reducing their exposure to climate conditions on extreme days. A silver lining of the covid-19 pandemic is that this drone delivery vision is accelerating as more companies work on this technology.[34]

Food Consumption and Climate Change

In 2014, the average American family spent 10 percent of its income on food.[35] Richer people spend a smaller share of their income on food than poorer people. This means that even large food price increases have only small impacts on their purchasing power. Given that poorer people spend a larger share of their income on food, a fair society will monitor food prices for price spikes on key food staples. If consumers purchase only locally grown food, then climate change could significantly affect their diet. Low transportation costs within and across nations allows people to consume goods that are not produced locally. If California has a poor avocado harvest, Californians can get avocados by importing them from Mexico. Discerning food eaters may argue that local produce is of higher quality, but for most people the Mexican imports will be a close substitute for the California avocados.

Consider an extreme case of a California avocado consumer who is unable to buy this product anywhere. Will this person's diet suffer? Substitution possibilities will determine the answer to this key question. An avocado is a bundle of vitamins (such as C, E, and K) and nutrients. The avocado has a price, and this information allows a consumer to calculate its bang per buck. The rational consumer will compare the nutritional content and taste of this food relative to other foods. If due to a bad harvest the price of avocados increases relative to the price of carrots, then this provides an incentive for the person to substitute away from avocados and to eat more carrots. Such a substitution will be more likely to occur if the person has a taste for carrots so that they are a close substitute for avocados. This person would suffer more from the price increase for California avocados (caused by the heat and drought conditions) if the person loves this variety of avocados and there is no other food such as carrots that is a close substitute. While such people must exist, such an unwillingness to substitute seems highly unusual. For example, suppose that two apples have the same nutrients as one avocado. If a person is willing to substitute to eating apples, then this person can easily adapt to a climate shock that raises the price of avocados.

In this age of low transportation costs for shipping food, the ability to purchase food from other distant producers insulates local urban consumers from agricultural price spikes. In an economy where people in Los Angeles cannot buy from farmers in other states, horrible weather in California will sharply raise the price of fresh fruits and vegetables. In an economy where goods can be shipped across states and international borders, there is a profit opportunity for farmers in Mexico and elsewhere who have had a bumper crop to ship these goods to the hungry customers in Los Angeles. The invisible hand of capitalism nudges such Mexican farmers to trade with the residents of Los Angeles. This trade benefits both the buyer and the seller.

The Role of Induced Innovation in Helping People to Adapt

As millions of US households and billions of people around the world seek new adaptation solutions, there is a huge market for firms that can devise products that help people cope. Our set of feasible coping strategies increases over time as innovation takes place. The next Elon Musk must choose what problems she will solve. The profit motive helps her prioritize.

Kentucky Fried Chicken's effort to introduce a vegetarian-based food product that tastes just like fried chicken offers a key example.[36] I recently ate my first vegetarian Impossible Burger and found it to be quite tasty, with the flavor of sizzling meat. In both cases, a for-profit firm took the risky step of investing capital to design and market a new product. Such a company will take such a step only if it anticipates through market research that there is sufficient demand for such a product. If enough consumers seek to enjoy the great taste of meat while eating less of it, then the for-profit company has an incentive to create a new variety that unbundles the two attributes. Once Kentucky Fried Chicken sells its vegetarian fried "chicken," households can change their diet while still enjoying the taste that they are used to. If climate change makes it more costly to produce ground beef or chicken, these examples show how market innovation allows consumers to obtain their food without sacrificing its tastiness. In this case, for-profit firms (through pursuing their own self-interest) help consumers to adapt to the rising cost of raising chickens and cows for meat production.

Market demand drives the direction of innovation. Drug companies focus their efforts developing new drugs for diseases featuring a large market demand for cures.[37] If few people suffer from a disease, then drug companies have much weaker incentives to pay the fixed costs and bear the risk of developing new medicines for that disease. Perhaps ironically, if climate change is expected to cause significant aggregate damage to the population through increasingly intense heat waves, then this creates a profitable market niche for those entrepreneurs who can devise solutions. For those new market products that require larger upfront fixed costs to design them, the for-profit firms will need to be convinced that the aggregate market is large. If the aggregate demand for such a heat solution is small and if the underlying research is risky to conduct, a risk-neutral firm may choose to delay making such an irreversible investment. In this sense, when entrepreneurs anticipate that there will be large future demand for an innovative climate adaptation solution, this creates the incentive for them to enter this market to design the new product.

This suggests that in the case of adapting to climate change there is strength in numbers. If you are the only household to experience mold due to heavy rains, then no firm will develop a new solution to remove the mold. If millions of people face this challenge because of climate change–induced heavier rains, then climate change induces innovation by creating enough anticipated suffering that it is profitable for firms to engage in the risky and costly search for solutions. The flooding caused by Houston's Hurricane Harvey created significant mold damage for thousands of homes. If some entrepreneurs anticipate these emerging opportunities, then the power of human ingenuity is unleashed in the search for solutions. Through a messy trial and error process, some of these new ideas will offer market solutions.

Entrepreneurial activity geared toward adaptation will be less likely to occur if large segments of the population do not seek out such products because they view them as a waste of money. If there are large numbers of climate skeptics, then this could inhibit technological innovation by reducing aggregate demand for these products. If aggregate demand for adaptation-friendly products is low and there are high costs to develop them, then entrepreneurs will be less likely to enter these markets.[38] If there are sufficient numbers of people who worry about the challenge of climate change and thus demand market products that help them to cope, then current climate skeptics benefit because they will have the option in the future to buy the new products that facilitate adaptation. Economics predicts that these individuals will respond to the law of demand. If the price of adaptation-friendly goods declines, because of economies of scale and because of global supply chains lowering the average cost, then they will be more likely to adopt them.

This logic suggests that the presence of a large number of climate deniers can slow down the overall economy's adaptation by stifling adaptation innovation.[39] Consider the following example. Suppose that: (1) the economy consists of one hundred people, half of whom ignore the climate change challenge; (2) the main climate change challenge is death from extreme heat; and (3) the fifty climate realists are each willing to pay $1,000 for an excellent air conditioner. Under these assumptions, the most revenue that an air conditioning producer could collect is $50,000. If it costs this producer $70,000 to develop and produce the air conditioner, then the producer will not do so because it would lose $20,000 in profit by developing it. Note that if all one hundred people become convinced of the climate challenge, then the producer would make the air conditioner because its revenue of $100,000 would exceed its costs of $70,000.

Given the significant degree of income inequality in the US economy, the super-rich play a special role in driving technological change. In June 2018, Bill Gates blogged about a new technology that guarantees that vaccines remain cold as they are transported to rural places.[40] He funded this innovation in part because he anticipated that there is a crucial need for making sure that vaccines do not spoil as they are transported to those who need them. Such cooling transport technology is another example of an innovation that fuels adaptation. Due to his personal fortune and his ambition to improve the world's quality of life, Gates personally helped to cause this innovation.

Development economists have been concerned that drug companies have much weaker incentives to design new drugs for people in poor nations in Africa because such drugs will generate less profit.[41] The same logic applies in induced innovation for mitigating climate risk. If billions of people will be facing similar challenges of extreme heat and sea level rise, new market opportunities will arise from this aggregate demand for solutions. In contrast, if hundreds of millions of poor people in a given area face a location-specific challenge, for-profit firms have less of an incentive to devote effort to innovate to solve this issue because aggregate demand for such a solution will generate less profit. In such a case, nonprofit foundations such as the Gates Foundation and other development agencies can play a role in promising a financial payout to the firm that successfully introduces a resilience solution.

As new goods are created and marketed, firms enter and compete against one another for market share. This competition benefits consumers as the price of such goods declines. Declining prices mean that even poorer people can afford these goods. The quality-adjusted price of key adaptation-friendly products ranging from air conditioning to cell phones to refrigeration have all declined sharply over time.[42]

The open research question here is how effective these new goods are in offsetting the new risks we face. Research documenting the role of the widespread diffusion of the air conditioner in attenuating the outdoor heat and death rate correlation offers a key optimistic example.[43] The widespread ownership of cell phones represents another example of a new technology that plays a key role in keeping people current on their social network, trusted news sources, and local government actions. This real-time information helps people make informed long-term choices and short-term decisions during a crisis. An optimist would note that even most poor people now have cell phones. In the next chapter, I discuss the quality of life challenges posed by climate change for the poor.

3

Protecting the Poor

Climate change compounds risks that poor people already face.[1] Poor people face a greater death risk from natural disasters because they live in riskier places in lower-quality structures and have less access to high-quality health care. Consider the case of Hurricane Maria and its impact on Puerto Rico when it struck in September 2017. As a widespread power blackout took place, poor, sick people in hospitals suffered interrupted medical care, and this contributed to the increase in the area's death rate. This example highlights how risks compound to increase harm. Sick people suffered more than healthy people from the macro shock (that Hurricane Maria struck) that injured key infrastructure, including roads, and access to electric power.

Climate change is likely to exacerbate the inequality challenge. Many Americans have little money in liquid savings accounts. When shocks such as illness, job loss, or natural disasters occur, the poor do not have access to savings to buffer their consumption during these times when they cannot work or have a sudden need to increase expenditures (such as to put down a month's rent deposit on a new apartment). After the November 2018 fires in California, some people were made homeless when their houses were destroyed. Poor people lack the resources to smooth their short-run transition as they try to rebuild their lives and find a new place to live. After major natural disasters, the federal government steps in and transfers a significant amount of money to affected areas. These funds cover payments for unemployment insurance and public medical expenses.[2]

In the midst of the covid-19 pandemic, the poor are suffering more as they live at higher population density, rely on public transit (exposing them to more contagion risk), and tend to work in service jobs that cannot be done at home.

The poor can protect themselves, and society can help the poor to cope with emerging risks, through a variety of strategies. I do not discuss the merits of federal government programs to reduce poverty here. Most economists support public policies that offer poor people income without reducing their incentive to work in the legal sector. The economist Milton Friedman endorsed the universal basic income as a policy tool to protect the poor.[3] Under such a program, people would be guaranteed a basic income level, and incentives to work and invest in human capital would be maintained. Some government transfer programs induce the unintended consequence of discouraging work and savings.[4] The future safety net will be stronger if these disincentives can be avoided.

Reducing Poverty Rates through Skill Acquisition

Acquiring human capital is the key to escaping poverty. Young children can begin to acquire such skill as their parents spend time nurturing them. For children from disadvantaged, single-parent families, society can augment such investments through providing early life enrichment. Through the efforts of parents, teachers, and extended families, as well as their own efforts, young people can acquire the human capital that allows them to solve problems, to be patient, to be able to imagine adverse future scenarios, and to be willing to experiment and try new things.[5] Each of these traits raises one's lifetime earnings and helps one prepare for risks in daily life.

An enormous education economics literature explores the challenges in delivering high-quality education for all children. The challenge that arises is that even the experts do not understand how different children learn. We know that parents, teachers, peers, curricula, and school resources (such as access to computers) all matter for children's intellectual and emotional growth, but there is significant variation in how these inputs affect the skill formation of any given child. The poor are more likely to attend disadvantaged schools with peers from a high-poverty background.

Such schools are more likely to face a challenge in recruiting and retaining high-quality teachers. Many poor children are brought up by a single mother and thus have had less parental access while at home relative to the average child.

In addition to these challenges, poor children grow up in more polluted areas, and these environmental effects slow their development. Because real estate is cheaper in less desirable areas, poor people tend to live there. In cities such as Baltimore, air pollution and lead exposure are both higher in the center-city neighborhoods featuring older housing. Exposure to lead paint, particulate matter, and polluted water all slow a child's development. Pregnant women are more likely to give birth to smaller, sicker babies when the mothers are exposed to more pollution.[6] If children are exposed to less pollution, then they are healthier and they learn more in school.

All of these factors compound and raise the risk that poor children will be poor when they are young adults. Facing these realities, a major research agenda in labor economics investigates the early life determinants of skill formation. Research on early childhood development emphasizes that grit and tenacity can be built up through early skill investment in the very young. If investments in early education help young children build up grit, then they will grow up and be more adaptable to challenges when they are young adults. Those young people who have more cognitive skill and grit will earn higher incomes and have a greater capacity to cope with change. From society's perspective, it is likely to be cost effective to provide early interventions for disadvantaged children than to wait and later have to spend government money on welfare and to house prisoners.[7] Labor economics research argues that, given the rise in computational power, labor markets will increasingly reward those with soft skills related to interacting with other people.[8] The early life intervention literature optimistically claims that noncognitive skills can be taught and nurtured.

Economists who work on the broad topics of the earnings of less educated workers and on public school education topics are making important contributions to climate change economics. If more poor people move up the income ladder, then they will have the skills and the financial resources to better protect themselves from emerging risks. Although the various policy levers that these scholars are considering are outside my scope here, this age of policy experimentation and evaluation raises the likelihood that cost-effective poverty-reduction interventions will be identified. The 2019 Nobel Prize in Economics was awarded to three development economists who use randomized field experiments to learn what works in reducing poverty in the developing world. Similar techniques are being piloted in America's cities to reduce urban poverty through interventions to help poor children receive a better education and to have labor market opportunities when they are young adults. While we do not yet know what works, there is a pathway to knowledge here that opens up the exciting possibility that poverty rates will fall in the future because of the new insights yielded by our experimental research culture.

Increasing the human capital of the workforce becomes even more important at a time when physical strength is becoming an increasingly obsolete skill. The decline in US manufacturing jobs has lowered wages for many adult men. China's rise as a global exporter has disrupted the economy in many of America's manufacturing cities.[9] If a local economy such as Pittsburgh experiences an unexpected reduction in local labor demand, then a whole cohort of middle-aged workers may be displaced. Such a person's family has planted roots in a specific neighborhood and city and thus faces high migration costs to move to a local labor market that offers more economic opportunities. Displaced manufacturing earners make less money when they transition to the service sector.[10] These service-sector workers earn less and thus have a reduced capacity to cope with new shocks.

Will New Poverty Zones Emerge in Increasingly Risky Places?

Areas such as parts of Kentucky that are repeatedly struck with climate shocks could grow poorer over time and lose population as residents move away and fewer people move there.

Poor people tend to live in the most risky parts of an area (in terms of crime and air pollution and disaster risk) because rents are lower there.[11]

With coauthors, I have published a research paper based on a data set that includes information on each US county for every decade from 1930 until 2010.[12] For each county, we collected data on the number of major natural disasters that the county was exposed to as well as the county's population level. We also collected data on real estate prices and county poverty rates. We found that in the aftermath of natural disasters, counties lose population, home prices fall, and the share of the population living in poverty increases.

Real estate lasts for decades. In cities such as Detroit where relatively little new housing is being built, much of the housing stock was constructed decades ago. Given a fixed housing supply, declines in local demand cause sharp reductions in local real estate prices. As real estate prices decline, poor renters benefit while people with better labor market opportunities in other cities will move out. As the share of the population who are poor increases, local tax revenue declines and the public schools become worse. Local retailers shut, and a type of local death spiral unfolds.

This dynamic has played out in cities such as Detroit that feature a durable housing stock and a declining demand to live there. When the vehicle industry boomed in the 1950s, Detroit boomed. Durable homes were built there. Due to international trade and the rise of the southern economy, Detroit lost its market share in producing cars, but the durable homes remained. As housing prices fell, Detroit became a poverty magnet, and the local tax base declined and public services suffered.[13] This chain of events has played out across the Rust Belt as factories closed down. Once an area becomes a poverty magnet, it is likely to suffer from rising crime and worse public schools. The perception that the area has low quality of life discourages outsiders from moving there.

A poverty magnet effect in disaster-prone areas will be less likely to occur if disaster recovery aid flows to places that have experienced significant disasters. Billions of dollars flowed to New Orleans after Hurricane Katrina and when Hurricane Harvey hit Houston. This rebuilding money helps these areas to stabilize and may actually create short-run booms as the areas rebuild. If disaster-prone areas are more likely both to be hit by shocks and to receive generous fiscal transfers, then it is less clear whether such areas will become poverty magnets because they are continuously being rebuilt.

Natural disaster exposure will not create a new poverty magnet if a geographic location has other attributes that offset the disaster disamenity. Suppose that a beautiful and productive city such as San Francisco suffers from a terrible disaster such as a flood or a huge fire. Such an event destroys many homes. The incumbent property owners face a decision whether to rebuild or to sell and leave the area. If richer people can build using more resilient materials and architectural designs, then the disaster may actually contribute to the gentrification of the area.

The town of Paradise, California, is recovering from the devastating fires of November 2018 that killed dozens of people and burned down many homes. Real estate sales are taking place there, and home prices are high.[14] New, more fire-resistant housing is being built, and costly fire-protection rules are being phased in. The major fire burned down overgrown trees that represented a fuel source for a fire. The new Paradise city that is emerging will feature a higher-quality, newer housing stock, and new rules that will protect it from the next big fire. The net effect of this dynamic is environmental gentrification. Poor incumbent residents of Paradise are likely to be priced out because they can no longer afford housing in their original community. The fire accelerated the turnover of the local real estate stock of homes. Poor people tend to live in older, lower-quality housing while richer people live in newer, higher-quality homes. Perhaps surprisingly, even though Paradise suffered such a destructive fire, richer people want to live there. This town is located roughly eighty-five miles from Sacramento and nestled near the Plumas and Tahoe National Forests. Such a location offers unique amenity features that other areas are unlikely to offer.

The fact that the Paradise fire is displacing poor residents highlights a key point that is stressed by urban economists. Public policies that are intended to protect a place (such as new fire code regulations) can displace people as market prices change. As a place such as Paradise becomes more climate resilient, its rents will rise due to the basic logic of supply and demand. The demand to live in such a place will increase. As rents rise, the incumbent poor who are less likely to own real estate there will be priced out. This process feeds on itself as more upscale restaurants and shops open up. An open question in the case of Paradise is whether the new fire regulations will be effective in offsetting the rising fire risk posed by greater heat and drought conditions. The gentrification of the area suggests that the people who are bidding for such new housing are optimistic about the town's future.

For the original residents of Paradise who are poor and can no longer afford the new rents to remain there, important questions arise: Where do they now move? How is their quality of life affected by this displacement? If they move to a more urban area and a poor neighborhood in that urban area, they may be exposed to less fire risk, but their children may now be exposed to more crime risk.

The Quality of Life Challenges Faced by the Homeless

The homeless face special challenges posed by natural disasters and extreme climate conditions. In California's major cities such as Los Angeles and San Francisco, the count of the homeless population has soared. This growth in homelessness is partially fueled by high local rents and by in-migration to the area. These cities have warm winters and generous social programs to help the homeless live.[15]

Urban economists argue that redistribution to the poor should take place at the national level to avoid the case where more generous jurisdictions (such as San Francisco) attract large numbers of homeless people to migrate there. As the homeless move from their place of origin to California, the taxpayers of San Francisco and California are punished for their generous benefits to the less fortunate. At the same time, society owes these individuals support because many of them cannot protect themselves and climate change further increases the risks they face.

In recent years, there has been an increase in heat deaths in the Southwest. In Maricopa County, Arizona, which includes Phoenix, reports the *New York Times,* "data compiled by the county's public health department show that the homeless represent a fast-growing share of heat deaths. In 2014, the county recorded seven homeless people who had died from heat-related causes. By last year [2018], that number had increased to 61 deaths, more than one-third of the total."[16]

Given that shelters provide protection against climate extremes, new research is needed to explain why some people choose to sleep outside rather than in a shelter. What upgrades of the shelter would entice more homeless people to live in such shelters? If homeless people stop sleeping on the streets and move into shelters, does their health improve? Does their exposure to heat and other risks decline? Are the homeless aware of the new risks they face? The rise of the big data revolution offers the potential for collecting novel new data. Suppose that a group of homeless people could be incentivized to wear smart watches to record where they are each minute of the day. Such real-time geocoded data would allow a research team to investigate such people's exposure to heat and air pollution and other risks throughout the day. Based on this information, those individuals who are exposed to the greatest risk could be identified and then be offered a tailored incentive program to encourage them to protect themselves. In this sense, the big data revolution facilitates increasing our understanding of who is at risk and what risks they face. This science-based approach helps cost effectively to protect the poor and helps them to become more resilient.

Can the Poor Afford Air Conditioning?

The poor tend to live in lower-quality housing in less desirable areas in older housing that tends not to have central air conditioning.[17] Such housing features cheaper rents. The US Department of Energy collects data on residential buildings' attributes. Based on the 2009 survey, the following facts were reported:

Central air conditioning and central space heating equipment often work in tandem or as a unified system. About 91% of homes built since 2000 have a main space heating system that includes central ducts; for homes built before 1940, that number is just 50%.

Although structural and geographic characteristics such as climate, housing type and ownership influence where air conditioning appears, access to air conditioning by low income households is much lower relative to other households.

Overall, 18 percent of households below the poverty line do not have any air conditioning equipment at all. About a third of households below the poverty line use room air conditioning compared to 15% of households with an income above $100,000. In contrast, about 75% of households with incomes above $100,000 use central air conditioning compared to just 44% of households below the

poverty line. The share of room air conditioners continues
to drop as more households, especially higher income and
owner occupied households, choose central air conditioning
equipment.[18]

A powerful room air conditioner can be purchased for $250.[19] This
upfront cost would be lower if this product could be leased on an annual
basis. The annual electricity consumption required to run this air conditioner will vary depending on local climate conditions. In Arizona, based
on data from 2009, the average household spent $1,600 on electricity each
year, and 25 percent of this was spent on air conditioning.[20] This means
that in this area, the average household is spending $400 a year on air
conditioning.

Many electric utilities offer the poor cheaper electricity. For example, in
Sacramento, the poor who qualify receive a 48 percent discount on their
electricity prices.[21] If poor people living in one-bedroom apartments in
Phoenix can rent an air conditioner for $50 per year, and if they qualify for
a 50 percent electricity price subsidy, then their annual air conditioning
expenditure will be roughly $250 per year. The philosopher John Rawls
stressed the importance of designing a society that protects its least well off.[22]

Poor people tend to live in the oldest housing in cities because old housing is of lower quality and features cheaper rents. Such older units tend to
have older appliances such as a refrigerator that is over twenty years old
and other energy-inefficient products. The new generation of durables
tends to be much more energy efficient (and thus cheaper to operate), but
older apartments feature older durables. The apartment building owner
has no incentive to install new energy-efficient durables if renters are not
willing to pay more for apartments that are more energy efficient. Given
the difficulty of estimating what your energy bill will be in an apartment
that you have never lived in, poor people are unlikely to factor into their
location decision what their energy bills will be in different apartments.
Whereas all new vehicles must post their fuel economy, only some cities
(such as Portland, Oregon) are now adopting legislation to require property owners to reveal their past energy bills to potential buyers and renters.
Even people above the poverty line, who must pay their own energy bills,
will not install heavy energy-efficient appliances into a rental apartment if

they do not expect to live there a long time. People who expect to live in an apartment for just two years will not buy an expensive, highly energy-efficient refrigerator because when they move they are unlikely to take it along. Thus, incumbent renters simply keep the old, inefficient durables in the apartment.

This means that the poor have less reliable and higher-operating-cost durables in their homes than richer people. This matters in terms of adaptation because this means that the poor have less reliable access to cooling services and refrigeration. Recognizing this problem, back in 2010, I proposed to the Los Angeles Department of Water and Power that it consider rolling out an offer to update energy-inefficient durables in older apartment buildings in the city. Under my proposal, tenants would receive a letter from the city inviting them to swap out their current energy-efficient durables in their apartments, and the department would install energy-efficient durables. The city would rent these durables to the tenants, and this rental price would be an extra item on the monthly electricity and water bill. My proposal would have been a win-win for the electric utility and for the poor people who live in the older apartments. The department could use its sheer size to negotiate a price discount on purchasing the durables from the manufacturers. It could hire installation experts to re-move the old durables and install the new durables. The city would gain from a reduction in aggregate energy demand (because these new durables are more efficient) because this would reduce its need to build costly additional power plants. The poor people would gain from having higher-quality durables. If these products are as energy efficient as advertised, their total energy bills (the rental fee for the new durables plus the total expenditure on water and electricity) could decline. I was unable to convince the department to pilot my idea.

Climate Change Will Exacerbate Environmental Justice Concerns

An important literature in environmental economics documents differential exposure to pollution and weather between rich people and poor people and among different racial groups. Blacks, on average, are exposed to more air pollution and toxic chemicals than whites.[23]

Richer people have access to more self-protection strategies. In early December 2019, the *New York Times* ran a front-page story asserting that richer people have greater access to shade in Los Angeles because trees are taller in their neighborhoods. The same article noted that bus stops are often not properly shaded.[24] Within a large city such as Los Angeles, when a heat wave hits, how much hotter is it where the poor live? In this big data age, this question can be answered by deploying low-cost sensors throughout the city to take the hourly temperature at many locations. Such real-time information would inform public policy targeted to protecting the poor.

In Los Angeles, the city opens up cooling centers on hot days.[25] Many poor people do not have cars, and public transit can take a long time to go from an origin to such a center as people must take several buses. One possibility is for ride-sharing firms such as Uber and public transit to partner on extremely hot days to facilitate the logistics of helping poor people access public services at a lower time cost. Such public-private partnerships facilitate adaptation. The ride-sharing firm scores political points with the city (that regulates it), and the city can redirect its less used bus lines to routes that are in higher demand. The net effect is a more cost-effective transportation network that delivers more people to cooling centers faster and thus protect them.

In my research with coauthors on the relation between crime and extreme heat and pollution, we find using eight years of data from Los Angeles that there is more crime on hotter days and that this relation is accentuated in the poorest parts of these cities that feature the oldest housing stock.[26] We conjecture that people who live in such areas have less access to air conditioning. On extremely hot days people without access to air conditioning may have more trouble engaging in self-control. Research investigating how to reduce urban crime documents that the risk of violent encounters among urban teenagers declines when these individuals participate in a training program that teaches them self-control techniques. This research finding supports the claim that people have hot and cold emotions such that they make better decisions (that is, ones that they are less likely to regret in the future) when they are less emotional.[27] If heat exposure raises one's irritability and anger, then violence can erupt during hot times. Teenage males in high-poverty areas are more at risk of engaging in violent crime activity. If extreme heat plays a role in provoking such individuals to engage in violence, then a long-run consequence of a short-run heat wave is potentially to provoke people into taking actions that impose costs on the victims and raise the likelihood of long-run incarceration risk for the perpetrator. In this sense, climate change increases inequality between the poor and the rich because the poor who live in these high-poverty areas face more risk on hot days and those who engage in violence will face long-run consequences for their short-run choices.

Geographic Mobility and Adaptation

In the United States, the poor are less likely to migrate than the average person. This matters because migration is a key adaptation strategy in that people tend to move to areas offering better labor market opportunities and higher-paying jobs. If the poor do not move, then they are more likely to remain poor because they are not matched with a local labor market that demands their skills. Richer, more educated people have the resources to finance their moves.

Poor people could improve their quality of life if they could be incentivized to move. The US Department of Housing and Urban Development created a unique field experiment called Move to Opportunity in several cities. Poor households randomly assigned to the treatment group received a housing voucher to pay their rent but were required to move to a lower-poverty neighborhood. This randomized research design has allowed researchers to estimate the benefits to young children from growing up in lower-poverty areas. Relative to an observationally similar control group, children who grow up in better neighborhoods achieve more emotional stability and higher later life earnings.[28]

This research suggests that an expanded housing voucher program that offers poor people the opportunity to receive rent discounts in areas featuring a lower percentage of poor neighbors and objectively less risk in terms of climate risk and crime could significantly improve the quality of life for the urban poor.

Rent affordability plays a key role in determining whether a family moves to a better area, but research also documents that personality traits such as being open-minded and having a willingness to try new things also correlates with migrating.[29] Migration represents sacrificing a sure thing (one's current place) for an unknown gamble. Those who are open-minded will be more likely to try out new experiences.

Economists have advocated for the use of housing vouchers (rather than subsidized public housing) to enhance the menu of possible housing units for poor households to choose among.[30] Many poor people live in subsidized public housing. The government chooses where to build this housing, and this conscribes the geography of where poor people live. The federal government's public housing program contributes to creating geographic areas with high poverty levels. The government has built housing and rented it out at low prices. Although the poor are attracted by these cheap rents, this means that their children grow up surrounded by very poor peers. Using a natural experiment set in Chicago in which public housing was demolished and its residents thus had to move to lower-poverty areas, researchers found that the children who moved to these lower-poverty areas were less likely to be high school dropouts later in life relative to their peers who lived nearby and remained in high-poverty areas.[31] This research supports the claim that growing up in public housing has scarring effects on youth. In the case of public housing in Houston, such housing also exposes its residents to more risk. In April 2019, the *New York Times* reported that Houston residents in public housing were facing more flood risk due to both where the public housing was built and the actual physical structures.[32]

Poor people may be less likely to migrate in order to keep their access to their existing social network. If a poor person is unemployed but has many friends living nearby, this may reduce the person's willingness to move to another city. The individual will recognize that he or she would have a better chance to find a higher-paying job in the destination area but would sacrifice access to friends. One study documented that those displaced from New Orleans by Hurricane Katrina moved to Houston and subsequently earned more money.[33] This suggests that New Orleans residents had built up social networks that anchored them to a less productive place.

If the poor have close ties to friends in their area, then these social networks will reduce the likelihood that they will be willing to leave a place even if it grows riskier over time. While social networks raise migration costs, they also offer insurance benefits for those who remain in a place. Social capital fuels adaptation when people can reach out to friends and family as a source of solace and financing during times of crisis. Altruism for friends and families creates a support network for individuals to work together during a natural disaster.

Some poor people have limited physical mobility and are confined to wheelchairs. Studies highlight the challenges this group faces during a flood. Such events only compound the challenges these disadvantaged individuals face. While we might hope that some talented engineer would devise an affordable wheelchair that can handle extreme flood conditions, in reality engineers tend to focus their efforts on profitable induced innovation. If millions of people were confined to wheelchairs, then there would be a large enough market to attract entrepreneurs to work on these problems. Sufficient aggregate demand for adaptation solutions triggers innovation. Given that innovation is costly and risky, entrepreneurs will be less likely to work on challenges that are endured mainly by the poor because their aggregate demand is not sufficiently high for suppliers to focus on these challenges.

Poor People in the Developing World

Outside the United States, billions of poor people live in rural areas throughout the world. One adaptation strategy is to encourage these individuals to leave the countryside (where the economy is more heavily affected by climate conditions) to move to the cities.

For young, less educated rural people, they will be more likely to find gainful employment in cities if the cities are industrializing. Manufacturing is a sector whose output is less likely to be positively correlated with agricultural production. Research on African urbanization find that 75 percent of studied cities specialize in agricultural middleman services.[34] This means that there is a positive correlation in income between African nations' rural and urban areas. Thus, when the rural countryside is suffering from bad harvests and the farmers seek out alternative sources of income, the majority of the cities are also suffering from a recession. The earnings gains from urbanizing are larger if the city offers manufacturing employment.

The ability to migrate within a nation hinges on its geography. Poor Pacific Island states such as Fiji, Kiribati, the Marshall Islands, Micronesia, Palau, Papua New Guinea, Samoa, the Solomon Islands, Timor-Leste, Tonga, Tuvalu, and Vanuatu all face natural disaster risk and have limited locations within their countries to which people can move.

The ability to migrate also hinges on the ability to afford the transportation costs to move to the city. A randomized field experiment in Bangladesh documents that financing constraints limit beneficial migration.[35] During this nation's monsoon season, farming output can decline sharply, and this threatens to depress poor people's consumption. Such individuals will be able to eat more if they or their family members can migrate to a city and work there. In this field experiment, a research team handed out at random bus passes that allowed the rural poor to migrate to the city to earn money. The experiment documented that this seasonal migration allowed the people in the treatment group to greatly improve their short-term consumption. This finding is highly optimistic because it suggests that improvements in transportation networks can facilitate rural climate risk adaptation. By sending a family member to the city, the family is holding a diversified portfolio such that it is earning income from the city and the countryside, and this helps the family to have more stable consumption. This result has also been found in the case of China.[36]

The Consequences of Rapid Urbanization

Severe rural climate shocks in poor nations, especially if they are expected to persist for years, will accelerate rural-to-urban migration. The rural poor are seeking stable jobs and improved lives. Access to improved medical care and better foods will help the urban poor to cope with climate risk.

The migration of more rural people to the cities lowers low-skill wages and bids up housing prices in these cities. Urban incumbents who compete with these rural entrants for jobs and apartments will see their purchasing power decline as their wages fall and their rents rise. The effects of migrants on local labor and real estate markets is hotly debated in the United States. The Mariel boatlift of Cuban refugees in 1980 represents a key case study of how cities can be affected by rapid migration.[37] Economic logic predicts that a rapid increase in the supply of low-skill workers lowers equilibrium wages for low-skill workers and raises local rents.[38]

In the developing world, poor migrants often move to informal urban slums. In these areas, people squat for years on land often owned but not policed or serviced by the state.[39] In such areas, rising population density can contribute to infectious disease risk and fears of violence. Residents in these areas live in low-quality housing such that natural disasters are more likely to cause greater risk of death, destruction, and disease. The slums tend to be located in the riskiest parts of the urban area, exposing residents to even more peril.

In these urban squatter areas, the local officials do not have legal jurisdiction. This means that the local leaders have weak incentives to provide public goods for these areas. If the formal city fears contagion due to the spread of infectious disease, this would create an incentive to provide public goods to the area. In the favelas of Brazil, local mayors may not connect the poor to the water and power grid because they fear that this expansion of services will have the unintended consequence of attracting more people to move to the area.[40]

Improvements over Time in the Poor's Resilience to Shocks

In May 2019, the *New York Times* reported the great news that the Indian state of Odisha suffered much less than what was expected when Cyclone Fani struck.[41] Information technology and the diffusion of trusted real-time information directed millions of people to safe locations. This optimistic example highlights how the widespread ownership and use of cell phones in developing countries helps people cope with new risks in real time.

A hallmark of capitalism is that the quality-adjusted price of goods continues to decline. As such prices decline for adaptation-friendly goods such as cell phones, refrigeration, and air conditioning, even poor people can afford these goods. Even the richest men in the past, such as John D. Rockefeller, did not have access to the technologies we use today. In this sense, market competition and world supply chains increasingly help the poor to access the same products that the middle class takes for granted.

This logic suggests that the price of climate adaptation–friendly products are declining over time due to competition and innovation. Climate change poses new risks to the poor's quality of life. For all individuals, the first priority is access to food. The poorest of the poor must be able to afford enough food to survive. They will face less risk of food price spikes if their nation's international trade policy features few barriers to importing foreign food. In this sense global free trade in agricultural products protects poor workers because it reduces the food price volatility they face. Since poor people spend a larger share of their income on food than richer people, they are more exposed to such potential price volatility.[42]

The poor will have greater real income if the price of key adaptation-friendly goods continues to decline over time. Free trade among nations facilitates the mass production of goods such as air conditioners. As the price of such durables declines, many more people can now afford these adaptation-friendly products. By operating these products, more poor people are protected from extreme heat. This logic chain highlights how international trade facilitates adaptation. Research based on long-run trends in the United States documents that the correlation between extreme heat and county death rates shrinks toward zero as air conditioning is more widely adopted.[43] This fact arises because over time a growing percentage of the poor has air conditioners.

Air conditioners are just one adaptation-friendly product that has grown cheaper over time. Declining prices means that the poor are more likely to purchase them. As the poor have greater access to self-protection devices ranging from cell phones to air conditioners to refrigerators, they will suffer less from climate shocks. To test this optimistic claim, a research team would need to identify a representative set of poor people in a given year (perhaps 1990), wait for a natural disaster to occur, and then resurvey this population to learn who died during this time interval and how they coped in the immediate aftermath of the shock. If this same research design could be repeated in the year 2020, the adaptation hypothesis posits that the poor are suffering less over time when a similar bad shock occurs because they now are better prepared to handle the shock.

The overall evidence from Cyclone Fani in India supports this claim. Additional household-level data need to be collected and analyzed to sharpen our understanding of how households are coping with similar shocks today relative to in the recent past. Do they now move to higher ground before the disaster strikes? Are their homes now sturdier? What exactly are their coping strategies? How effective are these strategies, and how much do they cost to implement? The answers to these microeconomic questions form the core for understanding how the poor cope with new risk.

4

Upgrading Public Infrastructure

On August 29, 2005, more than fifty sea walls and levees failed to protect New Orleans from the water surges associated with Hurricane Katrina. This contributed to a flooding of 80 percent of the city. Infrastructure failures like this could become more common in the face of more severe climate events.

Prudent public policy would seek to identify these weak spots and invest in infrastructure upgrades, but few elected officials prioritize the climate resilience agenda. These officials are tasked with addressing many different short-run issues. They must decide how much to spend on upgrading infrastructure that will benefit the area mainly over the longer term.

Urban officials in such cities as New York City, Boston, San Diego, and London have conducted prospective studies to pinpoint what likely challenges climate change will pose.[1] These predictions inform infrastructure investment decisions, but this information is unlikely to be sufficient to trigger increased investments. Although some cities, Boston among them, are now spending upward of $30 million a year on coastal resilience infrastructure, other cities are choosing to invest much less.[2]

In this chapter I examine the infrastructure decisions that cities, states, and the US government are deliberating. Despite the key role that public infrastructure plays in the local economy, city and state governments face tight budget constraints. During the covid-19 pandemic, cities and states face enormous budget deficits, pushing decision makers to reduce investments in core infrastructure upgrades and maintenance.

Such disinvestment helps to balance the budget but raises a jurisdiction's risk exposure.

The Public Infrastructure Investment Decision

We rely on public infrastructure to provide us with clean water, electricity, and transport services. Climate change poses new risks to such infrastructure. At any point in time, a city features the durable legacy of its past investments. New York City's century-old subway is a leading example, as is this metro area's aging John F. Kennedy Airport. Given that public infrastructure is long-lived capital, once it is built, the public decision maker faces an ongoing decision of whether to upgrade the capital stock. Deferred maintenance represents a gamble. The decision maker saves money in the short run but risks a significant future loss because the risk of ugly scenarios increases. For example, a city's older sewage treatment system will be more likely to overflow when heavy rains occur. In this case, raw sewage ends up on the streets.[3] Each city's leadership must decide what costs it is willing to incur to upgrade the existing capital and whether such upgrades are likely to sufficiently lower the risks of bad outcomes. In the case of reducing climate risk, predicting the benefits (before the project is implemented) is quite challenging. For example, how much risk will be avoided each year by elevating an airport's runways?

Given that elected leaders often face term limits, their short-term perspective on governing discourages them from making costly investments now that offer benefits only decades from now. Elected leaders are more likely to focus on immediate priorities such as reducing crime or upgrading public schools. Elected officials seek to be reelected, so if voters do not prioritize an issue, then elected officials are less likely to work to solve it. In this sense, a crisis (such as a crumbling bridge) creates a salient moment to upgrade infrastructure as voters focus on such an issue.

Infrastructure projects often feature large upfront costs.[4] A behavioral economics theory would posit that local officials are unlikely to prioritize upfront expenditures to upgrade infrastructure because the risk from not making such an investment is not salient. If a mayor invests $10 million in preventive expenditure, is she or he wasting money or a prudent risk avoider? If the mayor believes that voters are not interested in an issue until a crisis occurs, then before a crisis she or he has reelection incentives to underinvest in infrastructure. By this logic, a federal government that seeks to build up more local infrastructure resilience must nudge and incentivize public self-protection investment.

In the aftermath of a salient natural disaster, governments may then overrespond to signal to voters that they are doing something to address the problem.[5] A 2009 study documents that voters do not reward elected officials for proactive actions to mitigate threats but do reward officials who attract large federal emergency transfers to the region.[6] This suggests that elected officials who seek reelection have weak incentives to pursue costly resilience investments.

The Option Value of Delaying a Large Infrastructure Project

To protect the New York City region from flooding due to sea level rise and storm surges, the US Army Corps of Engineers is evaluating engineering and natural flood protection measures.[7] The *New York Times* reports that these investments could cost more than $100 billion and that there is considerable disagreement about which of these irreversible investments is most likely to deliver risk reduction for the region without triggering unintended consequences.[8]

When making an investment decision that features large upfront costs and significant uncertainty about the project's benefits, it can be wise to wait. As time passes, the menu of resilience investment alternatives will change. Ongoing progress in climate science will yield a more precise understanding concerning the actual likely risks that a specific geographic area will face. Ongoing progress in civil engineering will offer the investor a higher-quality piece of infrastructure at a lower cost.

The risk of delaying an infrastructure investment is that in the short run it exposes areas to the climate damage posed by the next storm. The risk of delaying a major resilience infrastructure project is smaller if construction teams can more quickly build the infrastructure once an investment decision is made. In US cities such as New York City and Boston, public infrastructure projects can take decades in part due to union delays, and in part due to lawsuits filed by various interests. In contrast, China built a six-thousand-mile high-speed rail network in less than a decade.[9] Institutions, rules, and regulations together determine how long it actually takes to construct infrastructure.

Diagnosing Weaknesses: Stress Tests for Public Infrastructure

In the aftermath of the 2008 global financial crisis, the US Federal Reserve evaluated commercial banks by conducting stress tests to determine which banks featured balance sheets that could withstand a major financial shock.[10] Public infrastructure could be subject to a similar set of stress tests. The goal of such tests would be to identify weaknesses in the existing infrastructure.

Ideally each jurisdiction's infrastructure would be audited by a trusted authority (perhaps the federal government) that would employ civil engineers to evaluate the future risks posed by climate change. If this information could be collected for each jurisdiction and posted to a webpage, then this would help to bring about accountability. Currently, the American Society of Civil Engineers issues state-level annual report cards on the nation's infrastructure quality.[11]

Reports about an area's infrastructure quality and risks will help bring about accountability as local residents and firms take notice. Elected officials would be asked what steps they are taking to protect the place. Resilience consulting firms and construction teams would foresee new opportunities in supplying solutions for areas with weak infrastructure. Ambitious firms would see that there is a growing demand for infrastructure solutions, and they would invest in the human capital to specialize in delivering these services. Given that nations around the world are wrestling with similar civil engineering challenges, the expertise gained in places such as the Netherlands can be used to help US cities to cope. As the economist Paul Romer notes, good ideas are public goods. Once we have such blueprints, they can be widely adopted if cities at risk seek out innovative solutions through a competitive call for project proposals. If these cities engage in nepotism and favoritism concerning who is awarded these infrastructure contracts, then the risk reduction benefits of the project will be lower and the cost of the project will be higher.

Financing Public Infrastructure Upgrades

If cities and states must use their own tax revenue to upgrade infrastructure, then their leaders will face different tradeoffs than if they can borrow from federal subsidized funds. If elected leaders face term limits, they may be unwilling either to raise taxes or to divert expenditure from schools or other valued public goods to less shiny repair and maintenance because future leaders will be the ones to gain from these investments. Elected officials prefer ribbon-cutting ceremonies where new pieces of capital such as a new airport are opened versus simply investing in long-overdue infrastructure maintenance. If real estate owners in an area anticipate that their property will lose value if there are infrastructure failures, such as breached levees, caused by natural disasters, then they form a powerful interest group lobbying local officials to upgrade infrastructure.

When cities can finance their infrastructure upgrades with federal subsidies, local taxpayers have weaker incentives to monitor the efficiency of the public construction sector.[12] A tradeoff emerges. Cities will implement more public infrastructure projects if the financing of these projects is subsidized, but their leaders will devote less effort to monitoring the costs of the project since they are spending other people's money. Given that it is difficult to measure the productivity of the public sector, taxpayers are less likely to receive a good rate of return on expenditures intended to upgrade the public capital stock.

Poorer cities have a smaller tax base because they do not feature booming high-tech industry or much desirable real estate. For these cities, the federal government could offer subsidized interest rates to encourage them to implement projects. If the federal government does not subsidize climate resilience financing, then spatial inequality will rise due to climate change. Poorer cities will have less financial capacity to finance infrastructure upgrades. Facing such financing challenges, the leaders of such a city will be more likely to delay investments and thus expose the local residents and firms to more risk. A richer city will be better able to finance such ex ante upgrades, and this reduces the location's risk exposure.

Landowners are a major beneficiary of improvements to the local quality of life. If new public infrastructure reduces local risks, then the local land becomes more valuable. Since landowners will gain the most from an upgrade in infrastructure, they should pay a larger share of the upfront costs of upgrading the infrastructure.[13] This logic suggests that by raising local property taxes, a city could collect extra revenue to pay for local defenses against climate risk. Higher local property taxes discourage incentives to invest in real estate capital because the local government will tax part of the value of the flow of earnings. If the extra tax revenue collected is earmarked for significant upgrades in local infrastructure, then even homeowners may support this tax because the net value of their asset would increase. The nuanced point here is that from the local property owner's perspective, a higher property tax bundles two different factors. Property owners lose from having to pay higher taxes but gain if the taxes reduce their properties' long-run risk exposure.

States such as California have enacted caps on incumbent homeowner taxes. Proposition 13 anchors the tax-assessed value of housing at the level at which the home was purchased. Given the real estate price appreciation that has taken place in California in recent decades, this sharply reduces the tax revenue that California cities collect from longtime homeowners, and in aggregate this reduces such cities' ability to invest in major projects.

An alternative to raising local taxes is for a city to issue bonds to finance infrastructure upgrades. During a time of ongoing low interest rates, cities must decide which infrastructure projects to prioritize. Baltimore is issuing bonds to upgrade the local public schools through the 21st Century Schools initiative.[14] Municipal bonds played a key role in the early decades of the twentieth century, allowing major American cities to finance their construction of water treatment systems.[15] Such investments sharply reduced mortality from infectious disease.[16] To encourage this approach, the federal government could offer a subsidy for bond issues focused on climate resilience, reducing the expense incurred by cities looking to invest in resiliency.

An innovative financing strategy is the introduction of performance bonds.[17] In the past, a bond buyer passively waited to be paid by the party that sold the bond. With this new financial instrument, the bond buyer has a stake in the outcome. The bondholder will be paid a higher interest rate if the project achieves its stated goal. Imagine a case where a small city seeks to invest in its climate resilience. If it finances this project with a performance bond, then those who hold these bonds have an incentive to lend their expertise to help the city to achieve its goals. Given the central role that human capital and expertise play in providing high-quality public goods, this alignment of incentives increases the likelihood that these projects succeed, and this facilitates resilience.[18] It is important to note that self-interest (not altruism) motivates the performance bondholder to take these actions.

Infrastructure pricing offers another source of revenue for cities. The public sector can use revenue collected from its infrastructure to finance public capital upgrades. People are accustomed to paying a charge for using a bridge such as the San Francisco Bay Bridge or a fixed cost for sewer fees to upgrade local systems. Such charges generate revenue that can be used to invest in maintenance and service upgrades.

Do Improvements in Public Infrastructure Reduce Risk Exposure?

An unintended consequence of upgrading public infrastructure in such places as flood-zone areas in Florida and Louisiana is that more people may choose to move to such areas. If this takes place, then the government's resilience investment actually increases the population's risk exposure. Governments need to anticipate how the population will respond to such infrastructure upgrades. The crowding-out hypothesis posits that when people believe that government investment has upgraded an area's safety, these individuals engage in less effort to protect themselves. For example, if people believe that an urban park is now safe at night because of an increased police presence, they may go on more night walks.

In the case of infrastructure, a tragedy could occur if people move to an area because they believe it is safe and their perception is based on a recent upgrade. This case would be more likely to occur if people placed too much trust in engineers. If the truth is that the area reinforced with new infrastructure still faces increased climate risk (due to rising greenhouse gas emissions), then people who fully trust the engineers to have climate-proofed the place unintentionally put themselves in danger. Those who know that they do not know the effectiveness of engineering methods in offsetting Mother Nature's punches would be less likely to move to such locations. Continued stress tests of new infrastructure conducted by impartial civil engineers could play a key role in measuring how important this issue is. The interesting social science concern that arises here relates to perception versus reality. If many people perceive the area as safe (either because of its track record in the recent past or due to trust in modern engineering) but in fact the area is at increased risk, then public infrastructure projects are more likely to backfire as people move to the hazardous area.

Upgrades in public infrastructure can be effective even if they displace private resilience efforts if the government has a cost advantage in supplying such services due to economies of scale. In Florida, as the government expands the sewage treatment capacity and the overall grid, fewer households need to invest in private goods such as septic tanks. Many coastal Florida homeowners will increasingly demand such sewer connections. Some homeowner septic tanks in coastal Florida are not functioning properly because of sea level rise. Given that private homeowners gain from public infrastructure upgrades of the sewage system, the sea level rise challenge should lead such homeowners to support a tax hike to pay for this infrastructure improvement.

In some cases public sector investments encourage the private sector to invest more in self-protection. Consider the case of fire risk in California:

> In California, some communities aggressively prepare residents for the eventuality of wildfire. Montecito, a community east of Santa Barbara, saw the need to address wildfire risk following the 1990 Painted Cave Fire nearby, which killed one person and consumed 427 homes. The Montecito Fire Protection District works with residents to reduce vegetative fuels along roadsides, create "fuel breaks"—essentially areas where native shrubs have been thinned or removed—at strategic locations on private property, and harden homes against embers by putting screens over vents and replacing siding and roofs with less flammable materials. Fire personnel help residents create "defensible space" around their homes by removing brush and dead trees. (As the name suggests, defensible space is an area where a home can be defended by firefighters.) The district also set up a neighborhood chipping program to help

residents dispose of excess logs and branches. It created a robust evacuation plan and educated residents on how it worked. And it changed certain codes, requiring new driveways to be wider and with enough turnaround space for large fire engines. Those things make it safer and easier for residents to evacuate and firefighters to get in to protect lives and homes.[19]

In November 2018, Paradise, California, made the national news for the terrible fire that burned down thousands of structures and killed almost a hundred people. Since then Paradise's officials have worked with the Paradise Ridge Fire Safe Council in educating the community on wildfire preparedness, increasing awareness of fire risks, reducing wildland fuel, and developing preparations for natural disasters.[20]

Such private-public partnerships foster resilience. The devastation suffered in Paradise raises several economic questions. For the residents, what were their perceptions of risk exposure before this horrible fire? Did this area attract risk lovers or people who cannot afford housing closer to Northern California cities? In the aftermath of the fire, are people moving from the area? For those who rebuild, what steps will they take to reduce their home's fire risk? What measures can the community take to upgrade transportation infrastructure and emergency evacuation plans to reduce the risk and the loss from future fires?

Enhancing Infrastructure Resilience

During natural disasters the electricity grid can be knocked out. In New York City, the utility Con Edison is working with Columbia University scientists and private consulting firms to plan out its resilience strategy to reduce its risk exposure.[21] How much climate risk can be avoided through such investments remains an open question. As Con Edison learns through a process of trial and error, other utilities can learn from its lessons.

Given the essential role of access to electricity in the modern economy, keeping the electricity grid running in the face of climate change poses new challenges. One way to raise the likelihood of constant access to reliable power is to build redundancy and surplus capacity into the grid. Such a costly plan represents a type of robust public investment because it lowers the probability of costly power blackouts.[22] A backup plan for the electricity grid includes investing in batteries and generators. An active engineering research agenda explores how microgrids and other decentralized approaches to power generation and distribution will perform under extreme climate events.[23]

After several major recent fires, the California utility company Pacific Gas and Electric (PG&E) faces liability challenges for underinvesting in power line safety. Solar and renewable companies are filling the void, signing private contracts to provide power through decentralized renewable power generation. If more power is generated locally, there will be fewer active transmission lines and a lower risk of future fires.

The California Public Utilities Commission faces calls to devote more attention to monitoring the quality, reliability, and risk of the state's existing power lines. Claims have been made that aging power lines have contributed to fire risk and that the commission has been slow both to hold the utilities accountable and to encourage them to invest more in precautionary upgrades of the infrastructure.

Whether fears of liability and lawsuits and increased regulatory oversight will nudge key managers of infrastructure to engage in new stress tests is an open question. At the same time that PG&E is being investigated for underinvesting in diagnosing and upgrading its infrastructure, San Diego Gas & Electric is evaluating the spatial locations of its major assets and whether these pieces of capital are likely to be in harm's way as sea level rise takes place. This utility's mapping exercise has revealed that climate change–driven flooding poses risks for its electric substations, transformers, and power lines. This information raises the likelihood that these key pieces of power infrastructure will be upgraded and protected. Such action will increase service reliability.[24]

Another infrastructure resilience challenge is upgrading cell phone service. The Federal Communications Commission is seeking out robust resilience strategies to guarantee reliable service during risky times.[25] Cell phone technology is vulnerable to disruption. If natural disasters damage delivery systems and key pieces of the network, such as cable sheathing and telephone poles and cell towers, then service outages will be more likely.[26]

Competition for Economic Activity

As extreme weather events occur over time, some cities will have a greater capacity to withstand such shocks. After Hurricane Sandy in 2012, the New York Stock Exchange was closed for two days. In the face of such dangerous events, if weather forecasters can offer a heads-up beforehand, then affected areas can shut down and reboot after the recovery. We will learn which cities are incapacitated by weather shocks and which are resilient. Those cities that gain a reputation (due to both natural geography and public investments) for having well-functioning infrastructure will have a competitive advantage in attracting footloose firms to locate there. If a city's infrastructure deteriorates, then local property owners will eventually suffer asset value losses because the city's productivity and quality of life will decline, which will reduce the aggregate demand to live and work there. Local property owners are a vested interest group with an incentive to push for upgrades of local infrastructure in the face of emerging climate risk.

The competition for jobs gives cities an incentive to compete for employers. One way is to offer companies large subsidies to locate there. In 2014 Tesla chose to open a major battery factory in Sparks, Nevada, near Reno.[27] The state offered Tesla more than $1 billion in incentives to locate the factory there. Economic research documents the local economic development benefits of attracting major factories.[28] An alternative to offering such tax incentives is to offer a great quality of life and productive, resilient infrastructure. Areas that offer such features will attract major employers.

Cities that suffer from low-quality public sector management will be at risk of losing valued employers and will be less likely to recruit the Amazons to locate there. Suppose a city has an airport that frequently delays flights due to mismanagement and a power grid that is often disrupted such that blackouts occur. Climate change will only increase this city's disruptions. Such low-quality public infrastructure will lead firms to leave the area, and new firms will be less likely to locate there.[29]

A given geographic area may be resilient in the face of climate risk either because it has good natural fundamentals or because of prudent local planning. Consider the major snowstorm that disrupted New York's

JFK Airport in January 2018. The airport's inability to divert planes arriving from Europe and logistical issues related to snow removal labor and capital led to chaos. After this event, the airport's directors and the federal government investigated what went wrong and what steps might be taken in the future to reduce the likelihood that this occurs again. The resulting LaHood Report indicates that JFK Airport will make reforms based on the hard lessons its leadership learned during this crisis.[30]

If a city underinvests in infrastructure and has trouble recovering from natural disasters, then this will lead to population and employment loss. This loss of the tax base will affect the city's balance sheets, and the rate of interest it must pay to borrow funds will increase. This logic suggests that local leaders will have an incentive to invest in infrastructure resilience if they anticipate that jobs and people are footloose and will be more likely to move to rival areas if local quality of life declines in the aftermath of major storms.

Although the conventional wisdom is that geographic areas that elect progressives to local and state offices will invest more in climate resilience, it is relevant to note that Republican areas tend to be right-to-work states featuring less support for private and public sector unions.[31] In such areas the cost of project construction and completion is lower. Facing binding budget constraints and lower costs for installing infrastructure, Republican areas may invest in *more* resilience projects because the price per unit of infrastructure is lower. This claim can be tested by studying whether geographic areas governed by Republican governors (there are relatively few Republican mayors of major cities) suffer greater or smaller economic losses when natural disasters take place than geographic areas governed by progressives.

As we have learned in cases ranging from the Boston Big Dig to the New York City Second Avenue Subway to the California Bullet Train, public projects can generate huge cost overruns if public sector unions use their clout to create a jobs program.[32] In Italy, Venice's ongoing attempts to build expensive antiflood structures highlight that many jurisdictions have trouble finishing public infrastructure projects on a timeline and within a fixed budget.[33]

Human Capital Builds Up Public Sector Productivity in Delivering Resilience

Major cities are investing in hiring management teams charged with preparing for anticipating new risks that the city may face. Such cities are creating offices of sustainability.

Consider this statement about the role that climate risk managers play in a mayor's office: "There are concrete ways that have proven successful in managing the risk of innovation. They include hiring someone institutionally in charge of innovation, like a chief innovation officer, and supplying that individual with a team of experts who understand how to deploy iterative processes for designing and trying new ideas and who can serve as internal consultants for all the departments in a city. These innovation managers should begin to think like financial portfolio managers in terms of managing risk."[34] This demand for public sector workers who are experts in resilience creates opportunities for workers who invest in specialized skills. This in turn creates incentives for universities to develop a curriculum to train the next generation of leaders. In this sense, specialized human capital is built up because cities commit to recruit such individuals to help them plan for the next crisis and for reducing the negative impacts of such disasters when they inevitably occur.

This discovery process of piloting new ideas and evaluating their effectiveness and then spreading the word about whether such efforts might succeed in other cities opens up exciting possibilities of reducing the cost of supplying public resilience. If one city demonstrates how to cost effectively upgrade its airport or how to reduce flood risk, then other cities' leaders can learn from such a guinea pig. An example is the effort in urban China to reduce flooding risk through introducing so-called sponge zones using green space to absorb water.

Public sector adaptation will be fueled by early movers who engage in resilience planning. The diffusion of these good ideas will be faster if city leaders participate in organizations featuring idea exchange such as 100 Resilient Cities. Such networks of officials meet often and exchange ideas and best practices about how to design resiliency solutions to the new challenges that cities now face. Note that this progress is grounded in human capital. Each city's officials know that they face new challenges and know that they do not know exactly how to address them effectively. Such officials seek to learn best practices from other cities. Through this process, good ideas spread: one city can adopt a similar infrastructure resilience strategy that has worked in another city.

This optimistic hypothesis concerning the diffusion of good ideas can be tested by examining whether smaller cities that do not have a chief sustainability officer adopt relatively inexpensive new ideas that have been proven to enhance resilience in larger cities. If such small and medium-sized cities do not adopt such ideas because they do not have a champion in the small local government, then this suggests that small cities will have greater challenges adapting than larger cities. In this case, the small and medium-sized cities would be the least prepared for new climate shocks due to both financing constraints and a lack of expertise in local government.

Los Angeles has committed to a goal of reducing its imports of water. Thus, the city's leadership has strong incentives to consider a wide range of solutions for enhancing its water supply.[35] Those who have innovative ideas will set up meetings with Los Angeles officials in order to earn large contracts to implement this new work. As the public sector in Los Angeles contracts with civil engineers to devise solutions tailored to the local goals, the lessons that are learned will be easily transferred to other cities that seek to emulate Los Angeles. This diffusion of good ideas embodies Paul Romer's vision. The key for good ideas to accelerate the rise of infrastructure resilience is that there be resilience guinea pigs—namely, cities that are willing to try unconventional solutions and to allocate the financial resources to compensate the engineers for investing their time and effort on the project. Once a good idea is discovered and piloted in the field, then this finding will become common knowledge through interactions and conferences and social media. Those local officials who seek out new ways to cope with an emerging risk can use web searches to quickly identify new strategies.

Entrepreneurs who are looking for new business opportunities and government contracts have strong incentives to stay in touch with such groups of city leaders because the for-profit firms will more quickly learn what possible public sector procurement contracts they might receive if they can devise a solution to an infrastructure challenge that a major city faces (such as upgrading JFK Airport). Economic logic predicts that the costs of supplying high-quality public infrastructure will decline over time due to the effects of learning by doing. As there is more global demand for public infrastructure resilience, more young people (engineers, architects, urban planners) will specialize their human capital and choose to enter this emerging field (of resilient infrastructure) and learn how to solve these problems.[36]

Disaster Planning

Throughout this chapter, I have focused on physical infrastructure capital such as power lines, sewer systems, and airports. The local government's human capital and disaster preparation help it to be ready for surprises. During and immediately after a major disaster such as Hurricane Sandy, a city's vulnerable population may be at risk of dying as nursing homes and hospitals lose access to power, food, and water. At such a time, disaster preparation and logistics plans and built-in redundancies in network systems (such as backup power generation) play a key role in protecting the vulnerable. During the covid-19 pandemic, we have learned that the federal government was not prepared to cope with a major national disaster.

In California, public officials are paid well, and these jobs have high status. With these incentives, government attracts and retains talent. Such investments pay off during a crisis. Consider the case of Paradise, which has faced fire risk for years:

> Residents trying to flee the 2008 fires were caught in massive traffic jams, flames burning on both sides of the road as they sat trapped in their cars. They clamored for local officials to come up with a plan. The solution created by Paradise city leaders was a plan that evacuated sections of the city at a time, said Phil John, chairman of the Paradise Ridge Fire Safe Council. They adopted protocols to convert two-way streets into one-way evacuation routes during times of crises. And some 70 people participated in a recent drill, rehearsing an evacuation down the town's main thoroughfare. All of this work "saved literally thousands of lives," John said. "There's no doubt in my mind."[37]

Disaster preparation represents a type of insurance policy. Where disasters occur, such prepared areas suffer less because key actors know what they need to do to protect the populace and plans are in place to not panic and to respond. The cumulative experience gained from coping with such disasters will help other areas to learn "what works" based on how different types of infrastructure perform during a crisis.

The challenge for a researcher intent on evaluating such planning's role in facilitating resilience is inferring the counterfactual. We observe the impact of Hurricane Sandy on Manhattan given the plans that Mayor Michael Bloomberg introduced, but we will never know how much extra damage this superstorm would have caused if Bloomberg had not invested time and effort into resilience planning. In the case of the covid-19 pandemic, how many fewer people in the United States in the year 2020 would have died from the virus if the nation had been better prepared? Since there is no valid control group, empirical researchers cannot answer this important question.

Federal, State, and Local Resilience Investment Coordination

Poor cities do not have the fiscal capacity to finance their own infrastructure and are likely to pay a higher interest rate in the municipal bond market. This means that poor cities, which are likely to have the lowest-quality infrastructure, will face challenges in financing upgrades of their infrastructure. As the New Orleans Katrina example highlights, this means that such cities are more likely to face catastrophic risk.

Although state government can be a source of financing and human capital, a political economy challenge arises. Many poor cities, including Baltimore and Detroit, feature mayors who are Democrats while the governors are often Republicans who look to the suburbs for voter support. In the 2018 governor election in Maryland, Republican Larry Hogan was re-elected with 57.7 percent of the vote. Only 15.7 percent of these votes came from the city of Baltimore. In Baltimore, he received 49.5 percent of the vote. The typical Republican governor who seeks reelection has an incentive to focus on her or his political base. This tends to be suburban voters.

In a setting featuring a poor progressive city and richer suburban areas, the Republican governor must tax the base supporters (the suburbanites) in order to collect the revenue to redistribute to the center city residents.

This discussion highlights that poor cities in such states will face an infrastructure financing challenge. Even in states with Democratic governors, these leaders are aware that the big city voters are unlikely to vote for the Republican. Economic logic predicts that state leaders will focus their attention on swing voters and areas where the electorate is on the fence concerning who they will support. Given that city voters tend to vote Democratic, the governor has an incentive to take them for granted.

At the presidential level, Republican administrations and Democratic administrations have different priorities with respect to investing in cities. Center city voters overwhelmingly vote for Democrats. This means that pro-city policies implicitly transfer money from suburban taxpayers to urban residents. Republican senators and a Republican president have incentives to oppose such transfers.

A further challenge in seeking Republican congressional financial support for urban resilience investment relates to the partisan nature of the climate change question. Although the late senator John McCain of Arizona argued that climate change poses a national risk and thus should not be a partisan issue, there is a clear left-right divide on the climate change mitigation question.[38] Given that Republican leaders in the House and Senate view climate change resilience as a Democratic issue, they have been less likely to work in a bipartisan way to further such legislation.

The Public-Private Productivity Interface

Senator Elizabeth Warren of Massachusetts famously said in 2011: "There is nobody in this country who got rich on their own. Nobody. You built a factory out there—good for you. But I want to be clear. You moved your goods to market on roads the rest of us paid for."[39] Her statement captures the interplay between the private market economy and public infrastructure. Our nation's firms use the public roads and airports to ship goods to final consumers. These firms rely on the Internet and the electricity grid to stay connected to their millions of consumers. In this sense, the private sector's productivity hinges on the resilience of the public sector's infrastructure in the location where the firm is headquartered and where it ships its inputs from and the transport infrastructure it uses to ship its final goods.

In the next chapter I explore how firms will adapt to climate change risk. Even if a major firm such as Amazon anticipates all climate events before they occur, Amazon's operations are embedded in cities featuring public infrastructure such as highways, airports, and sewer systems. The productivity of our private sector economy depends on the ongoing productivity of our public infrastructure. During the covid-19 pandemic, the ability of the Internet to handle the surge in traffic is a tribute to the increased capacity of our key infrastructure to handle surprise demand shifts.[40] Engineers will learn valuable lessons based on studying how this key piece of public and private infrastructure has remained so reliable during this global health crisis.

5

Will Climate Change Threaten
Economic Productivity?

Preserving productivity growth is essential for our overall standard of living. The law of 72 teaches us that a nation whose per capita income grows at 3 percent a year enjoys a doubling of its standard of living in twenty-four years while a nation that has a 2 percent annual growth rate experiences a doubling of income in thirty-six years. Macroeconomic growth hinges on productivity growth. A nation's poverty rate declines when the macroeconomy grows.

Our economy features many sectors, but our most productive firms are concentrated in a handful of high-tech industries, and these industries tend to cluster in specific cities. Given the place-based threats that climate change poses, could our major firms and key productivity hubs such as Wall Street and Silicon Valley be significantly damaged by climate shocks?

The microeconomic theory of the firm offers a framework to answer these questions. Just as households have many protection strategies, firms have access to an even larger set. Firms choose where to locate, where to source their supply chains, what output to produce and of what quality, which workers to hire, and how to collect big data and use such data to motivate workers. Such firms also choose how to protect their workers from risks and whether they allow some workers to telecommute. Together these different decisions create a menu of resilience strategies.

In thinking about the economics of maintaining and enhancing economic productivity, we must consider both productive urban places and their firms. Urban economists wrestle with the question of why there are such large differences in productivity (measured as economic activity per square mile or economic activity per worker) across space. One theory is based on *selection* such that superstars (in terms of both ability and ambition) systematically move to specific cities such as San Francisco and New York City. The other theory is based on *treatment effects*. This theory posits that there are positive synergies of clustering talented and ambitious people and their firms in close physical proximity. In this case, the sum is greater than its parts. Some urbanists worry that these fragile ecosystems that have taken decades to form are now threatened by new climate risks.

Could Wall Street Leave Wall Street?

The major productivity hubs in the United States are located on the coasts in such cities as San Francisco, Seattle, New York City, and Boston.[1] High-technology firms and high human capital workers have located in these urban areas to create productive clusters. In cities featuring large numbers of skilled workers and firms that seek to hire them, firms can search for the right workers who supply the skills and contribute to the work culture that the firm is trying to build. All these workers know their own goals and what they seek from a firm in terms of its mix of salary, hours required, and career growth opportunities. In a smaller city with fewer firms, workers would not have such a menu to choose from.[2] Young workers can see if they fit in at a certain firm given its culture and upward mobility opportunities. In a city featuring many firms, these workers know they can move around to find the best match given their tastes and talents. This dynamic between workers and firms means that both will be attracted to cities with a reputation for being centers of such activity. A type of self-reinforcing process plays out as an area that is known for being a center for high tech attracts people and firms that seek to be part of this industry.

In 2012, I was called by a reporter who works for the *Economist*. The reporter was writing a story about the risks that climate change poses for coastal cities. I was asked, "Isn't it true that Wall Street is a major center of finance in the United States?" I replied, "Yes." "Is it true that sea level rise could flood Wall Street as Hurricane Sandy recently demonstrated?" I said, "Yes." "Thus, couldn't climate change by accelerating sea level rise decimate the US economy by destroying Wall Street?"[3] The reporter's argument was that damage to the key productive places could greatly injure our overall macroeconomy.

Building on the work of Gary Becker and Julian Simon on human capital, I reject this place-based theory of economic growth. The human capital approach argues that any geographic area that has a pool of talent (either by attracting it or growing it) will experience economic growth. In the modern economy, places are productive if skilled people and well-managed firms locate there. The great financial cluster of Wall Street would become a less productive place if ambitious, young finance workers stopped moving to New York. Consider an extreme example such that sea level rise threatens Wall Street. If Wall Street firms take no preventive actions and if sea level rise is severe, then we certainly could suffer a major productivity shock as Goldman Sachs and other key firms experience disruptions to their day-to-day work schedule. Given that Wall Street firms make their money from anticipating market trends, such firms have strong incentives to be aware of the new climate risks they face due to where they have located their key workers and assets.

The area known as Wall Street solves a cross-firm coordination problem. Independent firms in the finance industry want to locate near each other to trade and learn from one another. The Nobel Laureate Thomas Schelling explained this idea. Schelling offered the famous example of meeting a stranger in New York City. As one writer explained it: "Let's say you know you have to meet a stranger in your city sometime tomorrow. The only problem is that neither of you can communicate in advance about *when or where* to meet. Do you think you could still manage to show up at the same place and time? Thomas Schelling . . . found in an informal survey that many of his students tended toward the same answer when posed this question about New York City: they would wait under the clock in Grand Central Station at 12 noon, hoping their partners had the same idea."[4] Similar to strangers seeking to meet, proximity to Wall Street solves the coordination problem of the financial sector. Since financial firms have historically located near Wall Street, the rational strategy for a new firm that wants to be close to the action is to also locate near Wall Street.

If the place called Wall Street is threatened by sea level rise, major firms such as Goldman Sachs will anticipate this threat and will move to higher ground (perhaps the Connecticut suburbs). Such firms will have strong incentives to hire expert geographers and climate scientists to find geographic areas that face less future climate risk. Goldman Sachs is a multi-billion-dollar firm, and its executives are paid more when the firm is more profitable. Such individuals have strong incentives to investigate nascent risks and to protect the firm's reputation for quality service. If Goldman Sachs exits an increasingly risky southern Manhattan, then other firms will follow, so that the economic agglomeration re-forms in a safer, more resilient area. If this dynamic plays out, New York City loses a major employment center and apartment prices nearby will fall in value but there will be new wealth creation on relatively safer land in the same large metro area. This optimistic scenario implicitly assumes that current clusters of productive firms can coordinate together and quickly re-form on higher ground.

This process will play out through a trial-and-error learning process. Given the physical size of the United States, there are many places to build our productivity hubs given that cities take up little physical space. This means that there are many possible places to build our future cities if our current productivity hubs face significant climate risk. This transition would incur costs and time. A silver lining of forming the new productive cluster on higher ground is that this would provide a new opportunity to reconsider what economic activities should cluster together and thus the new productivity center may be even more productive than the original cluster.

Urban economists have relied on natural experiments to study how productivity is affected when a past major center of productivity must shrink. One example is the decentralization of manufacturing jobs from Seoul as the South Korean government feared an attack by North Korea. This migration provides a natural experiment to test how firm productivity is affected when firms move away from big cities.[5] In the case of South Korea, manufacturing firms that left the big city experienced a reduction in their productivity.

If productive firms currently located in southern Manhattan had to migrate to higher ground, there are many suburban locations where they could move. An optimistic example is provided by China's experience with creating industrial parks. In a coauthored study, I documented how major cities in China acquired farmland and then installed road and electricity infrastructure to create new industrial parks, such as Lingang in Shanghai. These parks offer deep rental discounts, reduced taxes, and lower operating costs to lure productive firms to locate there. Because these parks are vacant, potential tenants who would benefit from proximity to one another have incentives to co-locate there. The total economic activity generated within such parks is more than the sum of its parts because productive firms' workers learn more from closeness to other firms and these firms conserve on transportation costs of goods and workers. Such park construction further stimulates the local economy as real estate developers build new housing and retailers open upscale restaurants and stores near the new industrial park. To create such an industrial park requires an active government that can offer the incentives to purchase the land and then repurpose it for urban production. Cross-city research documents a positive impact on overall city growth.[6] Those cities that attract new industrial parks in China tend to grow faster than they would have if they had not built the park. The productivity of many of these new Chinese industrial parks suggests to me that firms can migrate and re-form new productive clusters on higher ground.

Climate Adaptation Lessons from the Amazon HQ2 Location Decision

We have already looked at a single well-established finance cluster, Wall Street. What about other clusters? Major corporations such as Apple and Google choose where to locate their employees. Amazon's decision on the location of its new HQ2 made the national news. This is a valuable case study in that it highlights the fundamental role of expectations in making a decision with long-run consequences.

Amazon seeks to maximize its stream of current and future profits and thus has a strong incentive to consider the expected benefits and costs of choosing a specific location. If the leaders of the firm know that they do not know what risks a city will face, they will consider worst-case scenarios. They will hire expert consultants to advise them on the opportunities and challenges and tax advantages associated with each location. Given that a headquarters siting decision is a long-run choice, the leadership team has strong incentives to form expectations concerning each location's future challenges. For publicly traded companies, the executives must report to a board of directors. Part of the job of this board is to ensure that the leadership makes prudent investment decisions. This system of oversight creates accountability in making major corporate decisions.

In choosing where to locate, the firm's leadership will trade off the natural advantages of a location (such as San Francisco's sheer beauty) against the rents they must pay to locate there and the wages they must pay to recruit talent there. These leaders will also factor in the risks associated with this location as well as the local and state taxes. By choosing a location, the firm exposes itself to such local risks as crime and pollution and such climate risks as extreme weather conditions and sea level rise. If a geographic area proves to be less productive in the face of climate change and unwilling to make investments to adapt, it will slowly lose its job base. While incumbent firms may remain there, other firms will now be less likely to locate there.

Firms such as Amazon are publicly traded and have shareholders. Such shareholders suffer an asset loss if the firm's expected profits decline. As shown by the investment success of buyout financiers such as Carl Icahn and T. Boone Pickens, large profits can be earned by identifying underperforming companies and purchasing the low-priced shares (because the firm's current and expected future profits are low) and then firing the current leadership and replacing them. This competition in the asset market incentivizes a firm's leadership to be aware of emerging risks. Firms facing these rules of the game have strong incentives to consider the tradeoffs associated with long-run decisions of locational choice and the design of their production establishments.

A firm's productivity will be at risk if its management is overconfident about its ability to predict the future. If such a firm relies on historical data in determining where to locate, this firm's leaders will be more likely to be blindsided by new climate risks because they do not anticipate these risks and thus underinvest in precautions. New research entities such as First Street Foundation are emerging to use available information to measure historical flood risk for different land parcels and to predict future flood risk.[7] Firms have strong incentives to engage in due diligence before making a costly location choice.

When a firm chooses a new production or headquarters location, it will attempt to predict what its short-term and long-term profitability will be in various locations. Such locations differ with respect to their rent per square foot, the tax breaks the firm will receive, and the area's quality of life. This last factor embodies many dimensions ranging from average temperature each month, natural disaster risk, restaurant quality and variety, crime, pollution, and local school quality. Amazon will value these factors because these attributes of a place determine whether Amazon can attract and retain talent without paying combat pay in terms of higher wages.

If a specific location has cheap housing prices but the skilled do not want to live there, then leading firms will not locate there because the wage premium they would have to pay to attract and retain workers would be too high.[8] The most productive firms will avoid areas with the lowest quality of life.

Given that Amazon expects to spend at least thirty years in a given location, it has the right incentives to consider and consult with experts about how the area's economy and quality of life are likely to evolve over time. For example, if Amazon believes that climate change will make Nashville 120 degrees Fahrenheit in summer by the year 2023, then Amazon will be less likely to choose this location relative to a similar city that will be more livable in the medium term. Amazon will be less likely to open a new, durable, and expensive headquarters in a location that is expected to face significant new risks. Such firms will be more likely to choose a location with good fundamentals and where the urban leaders are upgrading the power grid and transport infrastructure to protect the area from future anticipated threats.

When Amazon was considering the Queens, New York, location for the new Amazon HQ2, a concern arose regarding whether this coastal location might face significant future sea level rise risk.[9] A CEO as shrewd as Jeff Bezos must be aware of this risk. Such anticipated risk means that such increasingly risky real estate will sell at a discount. Capitalism allocates scarce assets to those who can use them most efficiently. Amazon must have had an architectural plan for handling this medium-term sea level rise risk. A new headquarters can be made more resilient both through private investments in architecture and through working with local and state officials on a private-public partnership to invest in different infrastructure to protect the new buildings that Amazon will build.

Manager Quality

Human capital plays an essential role in determining whether a firm's managers and personnel are nimble. Firms have strong incentives to continually engage in risk assessment concerning what new challenges they are likely to face in the short run and long run.

In planning for climate risk, firms will be more likely to take prudent steps if they are run by managers who are aware of the known unknowns they are now more likely to face in the future. Such managers will plan for worst-case scenarios. Overconfident managers will be less likely to engage in such planning. Recent work set in India has studied how different manufacturing firms cope with high levels of local air pollution. Those firms with higher-quality managers have more success coping with episodes of high pollution.[10] This research raises the question of how to boost manager quality.

The count of high-quality managers can increase over time through investment. Every young person chooses how much time and resources to invest in education. In later life, there are additional opportunities to invest in online classes and other low-cost skills training (think of the Khan Academy). Each of us chooses how much skill to invest in and what skills to specialize in. As ambitious managers anticipate that climate risks can impede profitability and productivity, more of them will have an incentive to invest in human capital that facilitates adaptation. Such capital is not excludable. If I invest in these skills, this does not preclude you from making similar investments.

Managers' abilities to plan and perform during bad times indicate their leadership abilities.[11] If there is a competitive market for managers, then low-quality managers will be more likely to be fired and replaced with other managers who have the skills and personality to navigate the new risks the firm faces. Such a competitive executive labor market means that the most productive firms will identify weak bosses within the organization and replace them. For publicly traded companies, the shareholders will earn a low rate of return on their investment if corporate managers underperform. The board of directors of a firm will be better able to identify underperforming CEOs in this age of big data. Data on employee complaints, final customer concerns, and missing out on production goals are all clear signs of a company in trouble.

Such measurements foster accountability, and this reduces the likelihood that a low-quality manager retains power during a time when the manager's mistakes are costing the firm more profit. To repeat this important point, the cost of a firm having bad managers goes up during turbulent times. This creates an incentive for firms to hire strong managers, and this fact creates an incentive for individuals to train in such skills that help them enter the pipeline to compete for these leadership slots. Business schools play a key role here in enhancing this pipeline. If the boards of major companies foresee that resilience during climate shocks is affecting profitability, then they will increasingly search for managers with these qualities. This will create incentives for ambitious managers who seek to be CEOs to acquire these skills through executive education and other training programs.

In cases of firms that have bad managers and suffer profit loss when climate shocks occur, what becomes of the rest of the workers in the firm who are fired because the firm has underperformed? To study this question requires individual-level data (such as annual tax returns) to trace the well-being of these individuals over time. Labor economics research on the earnings dynamics for factory workers who have lost good jobs concludes that middle-aged workers who lose a manufacturing job face significant challenges in finding an equal-paying job.[12] The decline in Rust Belt manufacturing had significant effects on real wages because displaced manufacturing workers suffered a large wage loss as they moved to low-skill service sector jobs.

Manager quality has been shown to be an important determinant of firm environmental performance. One study using British manufacturing data documents that firms with higher-quality managers are more energy efficient.[13] This suggests that well-managed firms will be more resilient in the face of new climate shocks. This claim could be tested by collecting data for firm productivity on different days over the year. If the research team then also collects data on local pollution and climate conditions on the same days, then the researchers could study whether firms with better managers perform better on more polluted and hotter days. Such research could also test for anticipation effects. If better managers are better able to anticipate upcoming bad weather days, they may choose to front-load production on good days and choose to reduce production on bad days to allow the workers to recover.

Larger firms with better managers and firms with deeper pockets for financing capital expenditure are likely to be more adept at managing all types of risks. In the modern economy, these risks include cybersecurity, product safety, worker safety, and climate risk protection. The best managers will lead firms that will grow over time. If such firms are publicly traded, then they will have shareholders and boards of directors who will expect that the managers will foresee new risks on the horizon. Such managers will be more likely to keep current on climate science and place-based risks that their firms face. They will be more likely to make investments to maintain their firms' productivity even when the firms face extreme weather conditions.

Due to improvements in information technology and access to data, the modern firm has an increased ability to monitor its activities in real time. If climate shocks disrupt any part of the production process, then the revenue for this firm will decrease. Computer records allow the management to observe how the inventory of parts is evolving over time. If a storm impedes such a firm's logistics of shipping parts, then the sooner the leadership knows this, they can adapt by purchasing parts from another provider. If extreme heat reduces one of their factories' output, the firm's leadership will see this unexpectedly low production and will contact the factory boss to find out why that factory underperformed. If the factory boss says that heat caused the low productivity, then the firm's leadership will consider installing temperature monitors on the factory floor and take other steps to cool the factory in order to return production to normal.

High-quality managers use real-time data collection to evaluate firm performance and how it varies as a function of external environmental conditions. Such quantitative techniques help the firm to learn about how resilient it is under different conditions. By observing how different weather conditions affect each unit's productivity, the firm's management can prepare to be more productive even in the face of harsh climate conditions. The best managers will be the most responsive to updating past plans. Lazier managers will passively stick to their old rules of thumb for doing business and will thus be less likely to be prepared for the new shocks the firm faces. Those firms with more skilled managers will have a greater likelihood of successfully navigating these challenges. Larger firms will have the scale and the financing to contract with the best engineering and consulting talent to help them solve the problem. Smaller firms do not have a scale of operations to justify hiring the best talent to address a new problem. Many firm-level adaptation strategies involve paying a fixed cost to hire consultants to take a fresh look at the firm's production process or to install a new piece of adaptation-friendly equipment.

High-quality managers will have the self-awareness and self-confidence to admit what they do and do not know. This creates a profit opportunity for a new field of climate resilience risk consultants to be hired to work on the new challenges that firms face. Those firms that are unable to adapt to emerging risks will lose profits and market share. In the competitive landscape of the global economy, these firms will be replaced by other firms that prove to be more nimble. Although this dynamic will punish the incumbent workers and shareholders of those firms that cannot adapt, profits and opportunities will rise for firms that do adapt. The rising recognition of the importance of increasing corporate resilience creates a new market for consultants who offer solutions to emerging challenges. This development of human capital specialized in solving resilience challenges accelerates the adaptation process and helps to preserve the economy's overall productivity.

Firm Competition Fuels Adaptation

In industries where government has enacted barriers to entry that make it costly for new firms to enter the industry, incumbent firms will earn higher profits than they would earn if they faced new entrants. In such industries, firms may be slower to respond to changing risk patterns because they face little risk of earning negative profits. In this sense, intense market competition weeds out firms that are less nimble in the face of changing market and environmental conditions.

Companies that are privately held are more likely to face challenges adapting to new challenges. These firms tend to be managed by the family who owns the firm. Without the discipline of having corporate shareholders, such companies can be run by following old rules of thumb. The family-held firm is implicitly sacrificing profit in order to provide the heirs with guaranteed employment. Such a firm limits its executive search to members of the family. If an unrelated person could be a great leader of the firm, then the firm is potentially sacrificing profit for its nepotism.

When firms are run without the corporate oversight instilled by an arm's length board of directors, managers are less likely to be prepared for new risks. A skilled board will ask the tough questions and hold the firm's managers accountable. This expectation gives the firm's managers an incentive to devote greater effort and to seek out the advice of experts to better manage the firm. This oversight can mitigate concerns voiced by behavioral economists that many managers are myopic and overconfident.

In their book *Mastering Catastrophic Risk,* Howard Kunreuther and Michael Useem argue that many companies feature leaders who engage in behavioral tendencies, such as intuitive thinking, rather than in deliberative thinking.[14] They argue that corporate leaders engage in availability and hindsight bias as they rely on their own personal experiences for making decisions and focus their attention on the most recent crisis. Bosses also underestimate new risks, are overconfident about their abilities, do not engage in robust decision making, and instead do not think about emerging dangerous risks. Finally, the authors argue, bosses engage in the status quo bias of following rules of thumb that have been profitable in the past.

Each of these claims merits future research. If the new generation of corporate leaders exhibits these traits, then our aggregate productivity will suffer in the face of new climate risks because we will not prepare. But it is no accident that Jeff Bezos runs one of the world's most profitable corporations. Free markets reward those who efficiently use scarce resources and seek out new opportunities and manage risk. The behavioral view of the modern firm is more likely to be right for privately held firms in mature industries. Such firms do not have checks and balances to encourage innovation and experimentation. They do not feature enough competition to keep executives on their toes.

There is a certain irony that *Mastering Catastrophic Risk* is written by two leading experts at the University of Pennsylvania's Wharton Business School. Wharton is one of the world's leading business schools. Human capital plays a central role in making us better problem solvers. This matters because to adapt to climate change, firm managers will confront new problems to solve. Modern management raises firm productivity.[15] Given that most firm executives earn an MBA, the curricula of MBA programs will play a key role in building corporate resilience. Although students will continue to study accounting and statistics, Wharton and other leading business schools can accelerate corporate risk adaptation if new courses are taught on risk detection and machine learning for detecting new patterns revealed by corporate real-time big data.

Labor Market Competition Protects Employees

Because workplace amenities are not taxed, employers have strong incentives to bundle amenities such as health insurance and generous lunches into the workplace to raise worker morale and quality of life. A firm such as Amazon has a strong self-interest to create a workforce environment where workers can be happy and productive. In San Francisco, the temperature and the city's natural beauty deliver this. A firm that is able to create a wonderful indoor environment could go to Dallas and purchase cheaper land and then build an indoor Xanadu for its employees.

Recent research on building environmental quality highlights the role of a good work environment for facilitating one's health in the face of increased outdoor risk. One prominent recent example is Apple's new green headquarters.[16] The new Apple campus was built after consulting with experts who optimized the airflow in Formula One race cars and is meant to be a building that keeps worker productivity and morale high. The unusual shape of the Apple HQ helps with natural cooling. The structure's canopy protects employees from the harsh California sun. The ventilation system guarantees comfortable clean air offering temperature control.[17] Apple does not employ many workers who have little formal education. More than 70 percent of Apple's workers have a college degree. Since the US tax code taxes earnings but not benefits, the most productive firms have strong incentives to offer excellent workplace amenities that include free food, gyms, and environmental amenities.

For-profit firms have strong incentives to compare how their profits would evolve with and without investments to offset heat and pollution. Air conditioning requires incurring fixed costs of installing the capital equipment and then variable costs to pay for the operating costs and the electricity. An air conditioned firm will have more productive workers and happier workers. Such workers will be willing to work there for a lower wage than they would seek if they worked in more unpleasant conditions.

Firms will be more likely to adopt air conditioning if they have to pay higher wages without air conditioning or if it is less likely to retain trained workers without air conditioning. When workers have many firms to choose among, firms must pay higher wages if the job is less pleasant.

Such workers have choice and will seek out other jobs that pay roughly the same amount and have better work conditions. Since job benefits such as air conditioning are not taxed (unlike wages), firms in high-tax nations have an incentive to provide these benefits because they can then pay lower wages. If firms can produce much more output when workers are comfortable, then this will provide an incentive for such firms to air condition while the less productive firms will be less likely to do so.[18]

This prediction has implications for overall inequality. If the most productive workers work for the most productive firms, then pay inequality will understate actual worker compensation because the higher-paying jobs will also feature great workplace amenities that will insulate their workers from the heat while less productive firms will not provide these job amenities. Given that many workers, especially less educated workers, do not work at the most productive firms, these workers will face more unemployment risk due to climate shocks (because their firm's managers will be less prepared for these shocks) and the day-to-day workplace amenities will be inferior to that of a firm such as Apple. These dynamics suggest that climate change will increase inequality across workers.

The most productive firms that form the cornerstone of the economy have strong incentives to insulate themselves from emerging climate risks. They have access to the highest-quality managers and the financial capital to pay for extensive self-protection pieces of capital. The least productive firms that hire the least productive workers will be less likely to adopt these technologies. Differential access to climate change protection technologies will mean that the inequality gap between high-skilled and low-skilled workers will increase. If the price of air conditioning equipment and electricity declines, then this pessimistic claim is less likely to be true.

Adaptation through Telecommuting

In early March 2020, every university in the United States closed down and went online to protect people from the covid-19 virus. Professors were able to give their lectures while at home using Zoom software, and students learned while online. Research documents that more educated people are more likely to work in occupations and industries that are conducive to such telecommuting.[19]

Many for-profit firms are now switching over to remote working and conferencing via the web. In many American cities, traffic congestion has meant that average commutes can take over an hour each day. Firms that allow workers to telecommute are implicitly giving workers the freedom to avoid such a loss. As firms learn how to create teams from remote locations, this opens up several adaptation possibilities. By unbundling one's place of work from one's place of residence, workers will have much more freedom to choose where they live based on criteria other than commuting. This will help them to adapt to those risks and opportunities that they prioritize.

Improvements in information technology such as teleconferencing allow for a physical separation of a firm's headquarters from its back offices and its production centers. A firm can keep a few workers in a coastal center city location and send factories and back offices to distant locations. If coastal cities are especially prone to risks, then production activity can move to safer locations.

The productivity gap induced by separating various functions of the same firm is likely to decline over time due to the improved ability to communicate by phone, email, and teleconferencing. These technologies allow for close coordination without physical proximity. If a firm can coordinate across its various functions without them being physically near each other, then this opens up many more possibilities of where they can locate. This increased menu of locations facilitates adaptation.

More jobs now feature work that does not require face-to-face contact.[20] Thus, people can be productive during extreme weather events that limit their ability to travel away from home. Future snowstorms in Chicago will cause less short-run disruption for worker productivity because workers will engage in remote work on such days.

While dentists and baristas must physically face clients and coffee drinkers, there are many jobs in which a worker does not have to interact face to face with other workers or customers. More firms are allowing workers to work from home. This conserves on the need to rent commercial real estate and reduces commute times. Since more workers do not require face-to-face meetings to be productive as a team, there are now an increasing number of opportunities to allow people to work from home.

A field experiment set in China offers a preview of a future featuring more working at home.[21] A Chinese firm operates a call center where workers are paid to make calls and are paid a commission when they succeed. The experiment created an incentive to encourage a subset of workers to work from home. Contrary to the standard claim that working from home encourages shirking, the study found that those who worked from home were happier and equally productive and saved time from avoiding commuting.

This study is relevant for climate adaptation because it suggests that jobs that traditionally have been viewed as needing to be headquartered (so the boss could cheaply monitor workers) can be decentralized to multiple locations. In a setting where output can be measured and pay is an increasing function of such effort, workers have the right incentives to devote effort even if their boss is not physically nearby.

Another example of adapting to weather conditions is provided by the 2018 American Economic Association meeting. The association holds its annual meeting in early January, often in cold cities such as Boston, Chicago, and Philadelphia. Hotel prices are low in these cities during the winter months, and the association seeks to minimize travel costs for job-hunting graduate students attending the conference. Young, newly minted PhD economists seeking a job as an assistant professor must face rounds of interviews in hotel rooms. Inevitably, there are years when there is terrible weather, leading to canceled flights. In January 2018, many UCLA economists did not fly to Philadelphia in part due to flight cancellations. Rather than canceling job interviews, the faculty used Skype to interview candidates, and the job market search continued untouched. This example has several implications. UCLA's interview strategy saved California taxpayers money by not flying the faculty to the meeting. Since there were no handshakes, the probability of infectious disease spreading declined. The chair of UCLA's Economics Department (my wife, Dora L. Costa) told me that the interviews by Skype worked out almost as well as face-to-face meetings. This instance highlights how Internet communication technology facilitates adaptation to climate shocks. As the need for face-to-face interaction declines, more workers will have greater flexibility in where they choose to live and how often they go to work.

Competition in Financial Markets Facilitates Adaptation

Productive firms need to hire great workers and to raise capital to implement new projects. The suppliers of capital (hedge funds and Wall Street) are not altruists. They seek out high-return and low-risk investment opportunities. Companies that receive favorable ratings from such analysts will be able to borrow at a lower interest rate and will raise more money when auctioning off shares. This dynamic raises the virtuous cycle that analysts and finance market participants can play a constructive role in facilitating corporate investment in climate resilience. If some managers are slow to address the resilience challenge, then the adults in the room will be the financial sector. An investor such as Carl Icahn will engage in short selling a stock if he believes that the corporation will underperform in the near future. The interaction between for-profit companies and the financial sector nudges such companies to be aware of what risks outside analysts foresee and to be proactive in mitigating these risks. After all, even a climate denier CEO of a public company will take steps to reduce perceived climate risks if the company's bond rating has been downgraded due to damage the firm has suffered from natural disasters.

The cost of borrowing hinges on the activities of the ratings agencies. Firms such as Moody's Investor Service and S&P Global Ratings rate companies in terms of their creditworthiness. If the rating agencies determine that a firm faces risk, then it will only be able to borrow at a higher interest rate. These rating agencies are investing more resources to determine what risks climate change poses for firms. In July 2019, Moody's purchased a majority stake of Four Twenty Seven, a firm devoted to using climate science to create climate risk scores. Mark Carney, the governor of the Bank of England, has argued that climate change poses a first-order medium-term risk to financial markets.[22] If the ratings agencies hold firms accountable for their emerging climate risks, then these firms will invest more in risk mitigation precautions.

Network Disruptions and Supply Chains

Supply chains play a key role in the modern economy because few companies produce all of the inputs they need to make final products. Instead, firms specialize at what they are good at and subcontract out to buy inputs from other firms that have an edge in producing those parts. Consider the production of a car. A car maker could produce each part in the vehicle and then assemble the final car. Or a car maker could specialize and make a few parts but then purchase key inputs such as tires from a specialist. This second approach should reduce the average cost of producing the final car, but it exposes the producer to supply chain risk. Shocks to input providers (such as the tire maker) can paralyze the car manufacturer because it cannot access key inputs to produce the final good.

The novel coronavirus pandemic offers a scary case study of the costs of international supply chain disruption. When the virus outbreak caused a lockdown in China, Apple and many other leading companies had to grapple with how to produce products such as the iPhone. Foxconn, Pegatron, and other companies that produce key components of the iPhone have thousands of workers located in China who were affected by the virus outbreak. In retrospect, Apple's choice to have key features of the product be customized for its product means that it cannot easily substitute to other makers of a comparable input. The benefit of such customization is that the input works really well with the final product. The cost of such customization is that the final product maker is solely reliant on this key input. In an economy facing more place-based risk (both from contagion risk and climate change), this is an increasingly risky bet.

From the perspective of climate change adaptation, a silver lining of this tragedy is that firms going forward will develop contingency plans for how to respond when place-based shocks occur. A contagion is one example, but severe natural disasters can also disrupt supply chains.

Researchers have studied the disruption caused to Japan's economy by the 2011 earthquake and tsunami. These terrible events represent a natural experiment for testing the interconnections among firms that trade with one another. In the case of the Japanese economy and the tsunami, firm productivity suffered even for firms located far from the geographic ground zero where the tsunami hit.[23] A domino effect was observed as key firms in the nation's supply chain suffered disruptions because they were located near the area where the tsunami hit. As their production was disrupted, firms that rely on the affected firms to provide key inputs lost output because they could not produce their final output. In this sense, the shock rippled to areas not physically affected by the shock because of the economy's interlinkages. Macroeconomists are fascinated by this domino effect because it means that shocks to one geographic area can amplify and hurt the entire economy. This network effect stands in contrast to standard neoclassical logic that a negative shock to one area cancels out as activity moves to another substitute location.

Climate change could increase the risk of these cascading shock dynamics. If firms are aware that they are dependent on major input providers and if they know that climate risk raises the likelihood of disruptive low-probability events, then these firms will seek out new resilience strategies. In this case, the anticipation of more pain induces greater effort that helps to reduce the damage caused by the emerging risk.

By having access to alternative input suppliers and alternative transportation routes for shipping goods, along with multiple sites for producing final goods, the probability that a spatial natural disaster severely lowers a firm's production declines. In 2011, Thailand was producing 40 percent of the world's hard disk drives for computers. A major flood disrupted the PC manufacturer Dell's operations. Surprised by the damage the flood caused for its supply chains, Dell began to track the business continuity readiness of its suppliers as it sought to identify each partner's risks. It then diversified its suppliers geographically and pinpointed management responsibility for business continuity both internally and with its suppliers. It required risk metrics with frequent updates from both its partners and its own division.[24]

If there are sufficient numbers of firms that seek resilience for their supply chains, this will create new markets for innovations such as drone technology for improving logistics. If land transportation routes are disrupted, then drone deliveries become even more valuable. Firms can choose from many other self-protection strategies. Some will establish redundancies in their supply chains so that they can purchase key inputs from spread-out input suppliers. This geographic dispersion represents a type of diversification that reduces the likelihood that the prudent firm faces a future network disruption. Firms that are aware of the significant uncertainty associated with climate risk will be more likely to invest in such implicit insurance.

Companies such as Amazon can ship from alternate warehouses if there is a shock such as flooding to the transportation network. Firms along a river can have backup plans using railroads if barges on the rivers become stuck. Firms can hold more output in inventory if they worry about short-term shocks that hinder just-in-time production.

After Hurricane Katrina, Walmart set up its emergency operations center to help it track its global supply chains during good and bad times. Such monitoring is now facilitated by progress in satellite coverage.[25] This company recognizes that people rely on its stores to help them cope during a disaster. As a corporate communication states:

> When disaster strikes a community, our 12,000 locations across the world—which are stocked with the kinds of products that are crucial in an emergency—become a valuable resource. We work to reopen our stores as quickly as possible, so that people can access the pharmacy, grocery and other essential departments. Often, we can begin distributing items to those in need while response agencies work to establish ongoing relief operations. To provide supplies as efficiently as possible, associates in our Emergency Operations Center collaborate with first responders, nonprofits such as the American Red Cross, and local, state and federal government agencies. We also support relief efforts with our associates' expertise in logistics and operations.[26]

Larger firms will invest more in preparation for such rare events. Since such companies operate in multiple markets, they will have greater experience in handling disaster situations, and their different branches offer a type of spatial diversification such that these firms can move output across geographic markets if local logistics or production is disrupted. In this sense, larger firms are likely to have an edge in coping with climate change than smaller firms, which have less cumulative experience and less access to finance, and are more likely to have a lower-quality manager running the firm.

This discussion highlights the *distributional consequences* of natural disasters on firm productivity. Firms that are unprepared for these contingencies will suffer large profit losses when extreme events occur. Although these firms lose, rival firms will gain market share if they are prepared for the event. In this sense, competition among firms to remain profitable even during bad times helps the overall economy to be resilient.

Maintaining Service Reliability

The major firms in our economy rely on intermediate input suppliers such as electricity providers and cell phone carriers to provide reliable, high-quality service. If these pieces of the grid shut down during fires and other natural disasters, then the productivity of our major firms will be affected. This domino chain logic suggests that it is crucial that the intermediate input providers build resilience and robustness into their business plans. Intuitively, this means that they will need redundant services and to prepare for worst-case scenarios. Such firms will be more likely to do so if they face service area competition. If cell phone service providers could lose business if they develop a reputation for not being reliable, then the firm will have strong incentives to invest.

Major cell phone companies are taking proactive steps to ensure that backup power generators are fueled and ready in case the grid fails for a while after a disaster. If major companies anticipate that the core infrastructure is unreliable, then they will have incentives to invest in precautions such as having their own backup power generation and communications systems that are resilient in order to keep their operations functioning when severe shocks occur. Such backup plans are costly, and the cost-minimizing firm will face a tradeoff between investing in such backstop technologies versus shopping for a service provider with an excellent reliability record.

The service providers, such as for-profit power companies, have strong incentives to be reliable and vigilant. In the aftermath of the November 2018 California fires, Pacific Gas and Electric's stock price fell by 25 percent as it became public knowledge that its infrastructure may have contributed to the start of the fires. Facing this threat of increased liability risk, forward-looking companies in the service provision industry will have stronger incentives to self-protect. Such companies, as they pursue profit maximization in the face of climate change, will want both to continue to have a reputation for reliable service and to reduce their liability burden.

There can be a tension between reducing a service provider's liability to natural disaster risk and providing reliable service. An interesting case is playing out in California. During times of high winds in dry areas, there is an increased risk of fires starting if live power lines are knocked down. Anticipating this effect, PG&E responded in October 2019 by shutting off power. Hundreds of thousands of people and many firms in Northern California lacked access to power for up to several days to reduce wildfire risk.

This action offers the benefits of reduced fire risk but it imposes costs among those who now have unreliable service. PG&E's actions offer a test of how both households and firms adapt to blackout risk in which blackouts occur as a form of adaptation to reduce wildfire risk. I predict that more of these households and firms, now fully aware of the blackout risk, will take proactive steps such as purchasing backup generators to reduce the impact of the power outages on their quality of life and profitability.

Productivity Challenges Posed by Climate Change for the Outdoor Economy

This chapter has explored the microeconomics of how different types of indoor firms adapt to an emerging threat. Millions of workers work outside. On extremely hot and unpleasant days, what happens to these individuals? Who bears the cost of such work disruptions? In Los Angeles, many people hire gardeners, and there are many construction workers on sites. During extreme heat waves, these workers will only be paid if they do the work. They thus face a choice of whether to do the work and be exposed to the heat or to reschedule the work and suffer the income loss during these extremely unpleasant days. Amazon is selling cooling vests for $100.[27] How much protection these products provide to offset heat remains an open question.

There are smaller exurban communities whose members earn their livelihood from tourism and from natural resource extraction. Today, Maine is a center of lobster catching because the lobsters live in Maine's waters. Climate change is warming the oceans, and marine biologists predict that the lobster will migrate elsewhere.[28] In Maine's coastal towns such as Kennebunkport and Bar Harbor, some of the residents work in catching lobster. These individuals will lose their livelihood if the count of lobsters shrinks sharply. Already in Alaska, melting ice has shortened the crab season by a month.[29] The open question here focuses on how easily such individuals can find comparable work in the tourist sector in their hometown if their income from lobster catching declines. Resource extraction and tourist communities lack the job diversity for workers to transition easily into earning another source of income. If these people must move elsewhere to find work, what is their next best job opportunity? Will their quality of life drop sharply as they have to re-create their lives because they can no longer use their skills to harvest natural resources? The answers to these questions (which vary from person to person) determine the costs people face in adapting to climate change.

6

Protecting Urban Real Estate

Across the United States, homes sell for widely different prices. According to Zillow, the median home in California in May 2018 sold for $543,000, while the median home in Michigan sold for $141,000.[1] The value of a piece of real estate depends on location, location, location. Real estate is more valuable in areas featuring a booming local economy such as Seattle. Within such a metropolitan area, the same home sells for a higher price when it is located in a nicer neighborhood featuring such amenities as good local schools, low crime, and proximity to natural beauty (water, parks) and to culture, shopping, and restaurants.

People have an incentive to consider with care the merits of investing in a given property. Forward-looking home buyers will recognize that if they own this home for decades, they face future risks. A home buyer makes a place-based bet about the future desirability of the home, its neighborhood, and the city where it is located. A home is a specific structure (for example, a four-bedroom house with a pool) located near a given labor market in a neighborhood whose attributes depend on other residents as well as on business and economic activities. The area's elected officials also play a role in determining quality of life through determining taxes, regulations, and decisions over such public goods as investments in public schools and police protection. Climate shocks represent another place-based risk that affects the value of real estate.

Real estate structures are very costly to move, and this exposes them to climate risks. As the threats of climate change manifest themselves, land in areas that prove to have an edge in protecting residents from these risks will sell for a price premium. Zillow's researchers have made scary predictions about the aggregate US real estate losses that might occur by the year 2100 due to sea level rise.[2] This prospective research predicts that hundreds of billions of dollars of real estate may be submerged during this century. The prediction is based on counting the number of homes located in areas forecast to face significant future sea level rise. Given that Zillow estimates the resale value of each home in the United States today, it can combine its price predictions with the count of total homes at climate risk to predict the total value of the real estate stock that is at climate risk.

Although I respect Zillow's efforts for predicting the future in the year 2100, such a crystal ball forecasting approach is subject to the Lucas critique. As weather patterns change, the real estate market will evolve through both demand side and supply side adjustments such that a more resilient real estate sector emerges. This chapter explores how new climate change risks affect the demand and supply of local real estate.

Real Estate Market Dynamics Caused by Climate Change "New News"

The price of an asset reflects both its current fundamentals and its expected future fundamentals. In the case of real estate, if you own a new home that will last for seventy years, then the value of this home will equal the discounted expected rental stream for this property over the next seventy years. Why? Suppose the home's price is less than the discounted expected rental stream for the next seventy years. In this case, there would be an arbitrage opportunity for an investor to purchase the home at a relatively low price and then turn around and rent it out for this higher seventy-year total rental revenue flow. Arbitragers would recognize this point and would bid up its price until the arbitrage opportunity vanishes.

This annual market rent for a specific piece of real estate hinges on its location, the risks it faces, and the opportunities it offers for the people who live there, such as access to jobs, amenities, and leisure and shopping opportunities, and its lack of disamenities, such as noise, pollution, risk, and extreme heat. If this rent is expected to decline in the future because of climate change, then today's price of housing will be lower because it will reflect this future lower rent that the owner could collect by renting it out in the future.

The fundamental quality of life of an area plays a key role in determining the value of its real estate. In this sense, the owners of real estate bear the burden of bad news and gain the asset appreciation from good news. Land prices reflect the new news of the specific challenges or opportunities that a given geographic area faces. Over time as new news arrives (related to the local economy and quality of life), rents will rise in areas where demand is rising faster than the housing supply. Rents will rise if the local economy booms or if a major quality of life challenge such as crime declines or local air and water quality improves.

Areas whose fundamentals worsen will suffer an asset value loss to compensate people who are considering buying the home for the new risks that have emerged. In US cities, home prices tend to be lower in the hotter parts of town, and poor people tend to live in these areas.[3] As climate change raises summer temperatures, economic logic predicts that rents will decline in the hottest parts of a city relative to the rents in the more temperate parts of the city. Rent declines in areas experiencing increased heat compensate renters for their increase in discomfort.

Lower rents permit such renters to afford adaptation products such as better air conditioning. Unlike renters, homeowners in such increasingly hot areas within a city will suffer a loss in wealth as their home's resale price will be lower than it would have been without climate change.

This example highlights the importance of distinguishing between renters and owners of real estate. Owners experience a positive wealth change when there is good news regarding their neighborhood's economic vitality and quality of life, but the reverse is also true. In the absence of rent control, renters will pay less in rent when an area's economic vitality or quality of life declines.

Given that every person must live somewhere, relative comparisons matter here. If Boston and New York City are the only locations to choose from and Boston's quality of life deteriorates by more than New York's does, then New York's rents will rise. The current owners of this asset will bear the brunt of this new news. Across the United States, demand for housing is rising fastest (think of San Francisco and Boston) for housing in areas where the supply of new housing is quite limited. In such areas, both topographical factors (such as mountains) and land use regulations limit the supply of new housing. The demand and the supply of housing together determine real estate price dynamics.

Several studies measure the price discount for housing located in geographic areas that face sea level rise risk. In a paper studying the New York City area housing market from 2003 to 2017, the authors use the 2012 event of Hurricane Sandy as a natural experiment.[4] They observe each home sale transaction, and for each case, they collect information on the sale price of the asset, the date of the sale, the geographic location of the home, and the home's physical attributes. Using information on the geography of flood zones, they evaluate whether each home is located in one of the two highest risk hurricane evacuation zones as defined by the city's emergency management department. These are the geographic areas most likely to flood. Asset buyers have strong incentives to learn about whether a home they are considering bidding for is located in one of these zones. The authors use statistical methods to test for how the price of a home of a given size, age, and structure varies across space and time. After Hurricane Sandy, a home in the riskiest flood zone sold for 8 percent less than the same quality of home located outside the flood zone. The authors argue that this discount reflects the fear of future flood risk in these high-risk zones.

A 2019 study of the housing market in England documents that although there is a price discount for homes for sale in areas that recently experienced a flood, this discount vanishes over time.[5] One interpretation of this finding is one of salience. A major flood may shock people and they may overreact and avoid such an area, but after a period of normalcy home buyers return and bid aggressively again for such housing. An alternative explanation for why the flood effect on real estate prices vanishes over time is that other factors, such as nearby job growth or increasingly valued local amenities, offset the flood risk's negative effect on prices.

This example highlights how behavioral economists and neoclassical economists debate over what model of our behavior best explains recent facts. Behavioral economists would have a cleaner test of their salience hypothesis if all other factors remained constant over time. Unlike experimental scientists, we cannot run this experiment. With the observational data that we can collect, economists often face the identification challenge that multiple theories can explain the same facts we document. Because climate change will unleash new shocks whose severity we have not experienced before, it offers a high-stakes laboratory for testing behavioral economics ideas.[6]

A study from Florida documents that homes sell for a discount when they face more coastal sea level rise risk.[7] In this study, researchers also collected apartment rental data and find that rents (which reflect payments to live in a given location for the next twelve months) are not lower in areas that face future sea level rise. Because home ownership entitles the owner to the future rental stream of a given property, future expected climate risk affects the home's price but not today's rental price. This findings suggests that real estate markets are forward looking. Investors are aware that local quality of life changes over time.

Housing Supply

The built supply of housing plays a key role in determining the market price of real estate. Suppose that a Seattle neighborhood features one thousand identical, highly sturdy homes for sale. Suppose that fifty thousand people are considering bidding for these homes and that each bidder differs with respect to their willingness to pay (measured in dollars) to live in one of these thousand Seattle homes. If we sort the fifty thousand people's valuation from highest to lowest, the price of this Seattle neighborhood's housing will be determined by the one thousandth highest bidder. This is the marginal price setter. If Seattle had three thousand homes for sale, then prices would adjust there so that supply equals demand and the three thousandth highest bidder would set the market price.

In this example, I rule out the possibility of real estate developers building more homes in Seattle. In reality, such developers will compare their cost of buying local land and constructing a home versus the revenue they can earn by selling an existing home. Returning to the example above, if a developer builds a single home, then there are now one thousand and one homes in Seattle, and the most the developer could sell this home for would be determined by the bidder in this society with the one-thousand-and-one-highest willingness to pay to live there. A given location such as Seattle is a bundle of place-based features. Some of these are good (for example, mild winters and natural beauty), others perhaps bad (for example, natural disaster risk). If some people sufficiently demand the good features, then they will continue to bid aggressively for a piece of real estate even if it becomes objectively riskier over time.

Disagreement over Real Estate Asset Values

Potential buyers of an asset often disagree about the asset's fundamentals. In recent years, Elon Musk has been bullish about his Tesla company's future while many short sellers believe that it is overvalued and will experience a stock price decline. These investors disagree about this asset's value. Similar disagreement about the value of an asset arise in the case of climate risk and valuing at-risk real estate assets. Given the uncertainty and ambiguity about climate change's severity and the timing of this risk, it is likely that the same asset may be valued very differently by different people.

Those investors who are pessimistic about the local real estate market face an asset investment challenge as they attempt to make money based on their views. Suppose you are sure that over the next two years homes in low-lying areas of Miami will decline in value by 20 percent. What asset trading strategy should you pursue? If you thought that home prices would rise, you would try to buy homes in the area. Unlike in the stock market, where you can short sell assets, it is more challenging to devise a strategy for making money when you expect that real estate prices will fall in a given market. This asymmetry means that pessimists cannot act on their beliefs in local real estate markets. This also means that optimists who seek to know the opinions of others do not have a way to learn cheaply about the beliefs of others who may be better informed than they are. In national markets, for example, the market for Tesla stock, pessimists about Elon Musk's company can short the stock. Publications such as the *Wall Street Journal* report online in real time about the quantity of current short sales, and this is an informative statistic that even bullish Tesla investors will want to know.

In the case of coastal real estate (at risk to flood) and real estate near fire zones, potential real estate buyers often disagree about specific place-based risks. In this case, the most optimistic buyers will be more likely to bid the most for these assets. They could subsequently regret this investment when unexpected extreme climate shocks materialize. In this case, the asset prices of such homes will decline as more and more people update their risk beliefs.[8]

Public opinion polls consistently indicate that Democrats are more likely to be concerned about the climate change challenge than Republicans.[9] If Republicans tend to believe that climate science overstates the risks such properties face, then Republicans will tend to win the bidding for real estate properties featuring a beautiful current view but perhaps facing medium-term climate risk.

During a time of deep political polarization in the United States, it is important to consider the possibility that many climate skeptics recognize that there is rising climate risk. Such individuals may privately be taking actions to reduce their risk exposure while publicly continuing to state that this issue is not a national policy priority. If Republicans and Democrats do sharply disagree about the risks that coastal real estate faces in the future, then one possibility is for Republican investors to purchase such assets and rent their housing to Democrats to live there in the short run. This example highlights that there is a separation between who owns the asset and who enjoys living there. Renting breaks the link between the two. Real estate markets allocate the asset to the investor who values the asset the most.

If potential investors in real estate recognize that they know that they do not know the risks that the real estate may face in the future, they can do more homework to research the investment. Real estate buyers who make a large down payment (perhaps 25 percent of the asset's price) to buy a property have an incentive to do their background research and purchase risk reports from entities such as Jupiter or First Street Foundation that sell pinpoint risk assessments.

If society is concerned that some people (the most optimistic about a place's risk) will be less likely to purchase these reports, a government information disclosure mandate could be introduced. Such a mandate could require the seller of a home to provide potential buyers of the property with a risk assessment report. This light regulatory approach requiring information transfer from the seller to each bidder would reduce the chance that a real estate buyer vastly overpays for a property because of ignorance about emerging risks. It is important to note that I am assuming that the report that the potential buyer would receive accurately reveals the risks the property faces.

Even armed with these climate risk reports, some investors will continue to bid aggressively for risky real estate. This group will include people with a personal history that ties them to the area or who have idiosyncratic tastes for the location. The aggressive bidders will also include the optimists who continue to believe that the risks have been overstated by the climate scientists. As spatial pinpoint risk maps become of higher quality and cheaper to access, more and more real estate investors will be aware of the known unknowns concerning a specific property's climate risks. They will bid less aggressively for such real estate. Those who do purchase the real estate in increasingly risky areas (featuring flood and fire risk) retain an investment option. They can save money by deferring house maintenance and allowing the property to depreciate in quality. This strategy saves the investor money in the short run, and the investor preserves the option to upgrade the structure in the future if engineering progress takes place.

In recent years, real estate has become an internationally traded asset. The Internet is used to convey information about homes for sale, and cheaper air travel allows people to travel more easily. International capital markets allow rich people in other nations to transfer their money across international borders. People in China are buying many homes in California and in Vancouver. Russian investors are investing in properties in New York City. As US real estate markets attract international buyers, the scope of the market is such that the most recent buyer (especially in expensive local markets) might be someone from abroad. The ultimate bearers of climate risk are those who choose to buy and hold these assets in their portfolios.

On a planet with 7.4 billion people, some people will value buying one of these homes perhaps because of the area's beauty or proximity to their job. Those who are more optimistic that sea level rise will not threaten the local housing will bid more aggressively for such homes. When owners and buyers of assets disagree about the likelihood of emerging risks, then there are gains to trade as the most pessimistic incumbent owners sell their homes to the most optimistic bidders. The prices that emerge from this bidding process are determined by these interactions between the sellers and the buyers.

This process of determining the prices of housing is crucial for many households' wealth because most Americans hold their wealth mainly in real estate. Rather than holding a diversified portfolio of assets, such households own a single house, and their housing wealth is determined by the current market value of their house net of their current mortgage owed to the lender.

The Role of Mortgage Finance in Determining Real Estate Prices

The vast majority of real estate buyers use a loan to finance the investment. If buyers use only their own funds to pay for an asset, then they bear the full risk of this investment. If the home's price rises, they are richer, and if the home's price falls over time, then they are poorer. Because people borrow to pay for housing, they have brought to the table an established entity, the bank, with ample experience in such transactions. In this sense, the mortgage lender is the adult in the room. Even if individual mortgage borrowers underestimate emerging climate risks, banks have incentives to research these new risks. The standard loan contract is a thirty-year mortgage. This means that once the bank loans the money to the borrower, it is exposed to climate risk for the next thirty years because the borrower always has an option to default on repaying the loan. Banks that anticipate that they are exposing themselves to such ambiguous risk have several strategies for reducing their risk exposure. In risky geographic areas or in areas where the bank is not sure about the emerging risk, the bank could either charge a higher rate of interest or offer the loan for a shorter term (a fifteen-year mortgage versus a thirty-year mortgage) or require that the borrower make a larger down payment. In this last case, the borrower would have more skin in the game and would be less likely to default on a mortgage. Charging different interest rates for loans in different geographic places (due to climate risk pricing) would raise regulatory concerns about redlining. Minority home buyers have faced situations where banks rationed credit or charged higher interest rates because they deemed minority neighborhoods as risky (perhaps because of crime or other local attributes). If minorities disproportionately bid for housing in climate-risky areas, then this same issue will arise again. If banks cannot engage in climate risk pricing, then borrowers in low-risk areas will implicitly subsidize those in higher-risk areas by borrowing at a higher interest rate.

The interest rate that banks charge borrowers plays a key role in determining how home buyers bid for housing. A person will bid more aggressively to purchase a house if the bank that lends the money is more generous with providing a larger loan at a lower interest rate. Suppose a couple identifies a home they want to bid for that has a sales price of $700,000. If a bank is willing to make a $400,000 loan at 4 percent versus a $600,000 loan at 3 percent, that affects how the couple bids.

Lenders determine what interest rate they will lend at and the size of the loan. In the face of emerging climate risk, borrowers are likely to lend less capital at any given interest rate. By requiring lenders to hold more housing equity (defined as the value of the home minus the total mortgage), the lender knows that the borrower is less likely to default on the loan even if a natural disaster at the level of Hurricane Sandy occurs.

Consider a case where a home sells for $800,000 and the bank requires the borrower to make a $400,000 down payment and the bank lends the buyer $400,000. If a natural disaster occurs and the home's value declines to $600,000, the homeowners would not have an incentive to strategically default because the owner's equity would equal $600,000 minus $400,000, which is $200,000. If such homeowners defaulted on the mortgage, they would lose $200,000.

Although this last example sketched a case of the lender offering the buyer less generous terms, the real-world institutional details offer a more nuanced case. In my research with Amine Ouazad, we study lender behavior after natural disasters. We document that lenders continue to make residential loans after major hurricanes hit geographic areas but that they choose to securitize more of the loans they originate. Under the current rules adopted by the government-sponsored enterprises (GSEs), these lenders do not receive a price discount when they sell loans from an area that faces greater climate risk.[10]

Imagine a different set of rules such that lenders anticipate that the GSEs will offer a lower price for loans in climate-risky areas. In this case, the lenders have stronger incentives to develop more in-house human capital in the field of climate research so that they can better screen loans. Borrowers could be offered a menu of interest rates so that those facing greater risk will be quoted a higher interest rate if they take no defensive steps and a lower interest rate if they invest in climate resilience. The key point to note is that the change in the GSEs' pricing strategy increases the likelihood that the lenders engage in this increasingly sophisticated behavior. Such lenders would have an increasing incentive to recognize that they are on the hook for the rising climate risk. Why? Homeowners whose property suffers from a natural disaster hold an option to default on their loan. If the property has been upgraded to better withstand an intense disaster, then the owners will be less likely to default. Recognizing this dynamic, the lenders now have an incentive to nudge the borrowers to upgrade the structure.

Financing Home Improvements

Homeowners may be financially constrained and thus unable to make profitable investments upgrading their homes. For example, imagine that it costs $120,000 to elevate a home to reduce flood risk. A household with only $35,000 in cash deposits at the bank cannot finance this. The bank would be willing to make a loan if the home is used as collateral, but this requirement is likely to diminish the willingness of the homeowners to make this investment. Middle-class and poorer homeowners will have less capacity to self-protect and thus will face greater losses if extreme climate events occur.

The market solution is for such homeowners to sell their homes to management companies. These managers could diversify risk by owning homes (that they rent to the current owners) in several locations. These profit-maximizing firms would be less likely to be financially constrained and would invest to upgrade the resilience of the homes. They would charge a higher rent for this improvement. If incumbent homeowners are underinvesting in resilience and if this feature is increasingly valued in real estate, then markets will create an orderly process to transfer these scarce assets to a party that has an edge (a comparative advantage) in cost-effectively upgrading the asset.

The Role of Real Estate Insurance Rates in Determining Real Estate Prices

If homeowners can buy cheap private sector insurance that covers various risks, or if the government offers subsidized insurance, then people will bid more aggressively for such real estate and its market price will be roughly similar to comparable properties in safer areas.

When asset owners are implicitly subsidized for bearing risk, the owner of the asset bears less risk. The US government, through FEMA, offers low-priced flood insurance. These subsidies mainly benefit wealthier real estate owners.[11] Congress has debated raising such rates to reflect the rising actuarial risks that such coastal properties now face.[12]

Climate change could raise the likelihood of eliminating such locational subsidies, and this would foster adaptation. Although more Americans now live in flood and fire zones (because these are high-amenity areas), most taxpayers do not live in such areas. As this majority of taxpayers learn that the subsidies they are paying to the risk takers are growing over time, reforms to phase out these policies become more likely. As the social cost of an inefficient policy grows, the political coalition with an incentive to push for reform is willing to devote more effort to counter status quo policies.

Under President Barack Obama's Affordable Care Act, there is a mandate that each person must buy health insurance. Suppose that each American real estate owner must buy at least a basic level of flood insurance. A benefit of such a government mandate is that this would create a large national flood insurance market. This would create an incentive for insurance companies to invest in talented analysts who know how to use spatial data to estimate the risk facing each property. Start-ups such as First Street Foundation are using historical data to calculate property flood risks.

In a competitive market for property insurance, those properties deemed to be riskier would be charged higher insurance prices. Market competition among insurers would help to reduce concerns about price gouging on such premiums. Insurers might offer households discounts if insurance buyers demonstrate that they have taken self-protection actions that reduce the property's risk exposure.[13] There are important synergies between a home buyer's insurance policy purchase and the homeowner's mortgage terms. If a home buyer purchases an insurance policy that covers climate contingencies (such as a Hurricane Sandy–level shock), then mortgage lenders will lend at a lower interest rate because they face less default risk if such a natural disaster occurs and destroys the borrower's home. An open question concerns whether there would be climate resilience gains if the same company issued both mortgages and home insurance policies because this would increase the likelihood that these two decisions are well coordinated.

Exposure to extreme financial risk also takes place for homes in the American West fire zone. As parts of California have faced greater fire risk, some insurers are sharply raising premiums on fire damage insurance or not offering insurance at all. If insurers exit the market, this means that homeowners in fire zones are exposed to large possible losses if their home burns down because they will not receive a dollar payoff if this event occurs.

In December 2019, the California insurance commissioner took the unusual step of preventing insurers from dropping policyholders from policies for those who live in fire zones.[14] This action is effectively a subsidy for those who live in these areas because they receive cheaper insurance than they would have received under a market-pricing mechanism that reflects the emerging risk. California insurance companies will turn around and raise the rates they charge other California home insurance policy buyers, so in effect those who live in relatively safe areas will be cross-subsidizing those who live in risky areas.

We have now explored how the markets for real estate finance and insurance play key roles in determining the market price of real estate assets and ultimately who pays for the financial risks caused by climate risks. To appreciate this point, consider an example such that the Smith family wants to buy a coastal home that is predicted to face greater sea level rise risk in the future. If lenders are aware of this future risk and they worry about extra mortgage default risk (and they cannot easily securitize the mortgage), then they will be willing to lend the Smiths less money and charge them a higher interest rate for the money they do lend. The Smiths will respond by bidding less aggressively for this home. The same logic applies for insurance. If private insurers charge the Smiths a high premium for a flood policy, then the Smiths will bid less for this increasingly risky home. In this case, the current owner of the increasingly risky home will suffer an asset value loss because potential buyers such as the Smiths will bid less aggressively for it.

Some people underestimate new emerging risks and bid aggressively for increasingly risky real estate. In such a case, the adults in the room can be the mortgage lender and the property insurer. If the federal government does not subsidize risk taking, these profit-maximizing firms have the right incentives to hire experts to assess risks and charge home buyers a higher interest rate for a loan and a higher insurance premium price for insurance if the home faces risk. These adults in the room can provide incentives even for behavioral real estate investors to take proactive steps (such as investing in stilts to elevate a home) to reduce its risk exposure. The net effect is a more resilient real estate stock.

Anticipating Threats to Real Estate Assets

In the greater Santa Barbara, California, area, a huge fire in December 2017 burned thousands of hillside acres. Following the fire, heavy rains triggered mudslides in Montecito that destroyed homes and killed people. This horrible case highlights the negative synergies caused by climate change. If climate change exacerbated the fire risk (due to drought and heat conditions) and if climate change contributed to the deluge of rain, then climate change is partially to blame for this disaster. Homeowners can take proactive steps to reduce the damage their property suffers from mudslides. These include landscaping and diverting mud flows away from the property. Following this salient disaster, homeowners are now more likely to take these costly steps.

If potential home buyers are increasingly aware of the rising risk of rare events, then this will create new demand for trusted information. This creates a new growth opportunity for consulting firms that engage in geospatial research and use climate modeling to predict risks for real estate property. In the case of coastal real estate, such companies already exist. A company called Coastal Risk Consulting generates home-specific indicators of flood risk. In a typical report, the company reports the home's FEMA flood zone status, predicted annual flood chance, local wind zone exposure, elevation and risk from tidal flooding, and risk of hurricane storm surges.[15] Such information enables home buyers to be more informed about the risks they face before they purchase an asset and what upgrades would further reduce the home's risks. "There's no one size fits all," says Albert Slap, president of Coastal Risk. "What one home needs may be very different than what the house right next door needs."[16] These reports help potential real estate asset buyers to foresee new risks—even if they have extremely low probabilities of taking place. Those potential buyers who are informed about these risks will shade down their bids for risky homes. This means that the household who actually buys the risky home is compensated for taking this gamble and the owner of the risky asset has an incentive to upgrade the defenses of the home in order to collect a higher home sale price.

How can an amateur (the home buyer) know that the supplier of the risk report is using the best climate model to estimate the risk that a given property faces? Reputational concerns matter here. If a company such as Coastal Risk, which claims that it can predict flooding, has a faulty model, then its predictions will be shown to be inaccurate, and the media will cover this and there will be litigation. Those companies that overstate the quality of their forecasts will be more likely to go bankrupt, and this will provide a strong incentive for such firms to invest money to improve its modeling capability.

Cross-City Real Estate Asset Competition

More and more economic activity is located in coastal counties.[17] The trillions of dollars invested in durable real estate means that investors are betting that coastal areas will continue to be productive and offer a high quality of life. Although deaths from natural disasters continue to decline in the United States, significant damage to physical assets occurs when storms such as Hurricane Sandy in 2012 hit.[18]

If sea level rise does submerge parts of Wall Street, then this part of downtown Manhattan will suffer a loss in real estate value while owners of real estate in nearby Greenwich, Connecticut, will gain as finance jobs will move there. In this sense, rising coastal sea level risk creates losers and winners.[19] Those who own the buildings and the nearby residential real estate in flood-prone areas will suffer a capital loss, but there will be capital gains in the new financial cluster where the finance industry might move (for example, Greenwich).

If Wall Street loses firms due to sea level rise, then Wall Street's landowners and the borough of Manhattan suffer. This economic activity does not vanish. Instead, it moves to a rival location. When nearby cities compete for residents and jobs, what matters is relative comparisons. If Boston suffers from more extreme heat each summer than its rivals, then businesses will be less likely to move to Boston. If climate risk increases in Lower Manhattan, then the demand for close substitute places such as Greenwich increases.

Given that the United States is such a large nation, if any one geographic area suffers from climate change, it is likely that this benefits a competing area. A particularly salient example is ski resorts. Skiing suffers when temperatures rise above freezing. This sharp discontinuity is being observed more often in ski resorts at lower latitudes. One research team geocoded ski resorts and documents that condo prices are declining in areas close to "muddy" ski resorts but are rising at ski resorts at northern latitudes.[20]

This research highlights that climate change creates geographic winners and losers such that relatively less risky areas will experience real estate price appreciation. The people who travel to the ski destinations will now have longer commutes (otherwise they would already be going to the northern resorts). This point matters because one might conclude that this is a zero-sum game such that the losses suffered by the ski resort that loses revenue (because the snow has melted) is just offset by the revenue gained by the newly desirable ski resorts. While the transfer of economic activity is not a zero-sum game, the popular media often focus solely on the losses incurred by the area that loses due to the climate change and ignore the harder-to-measure gains for the competing rivals that may be spread out and in several locations.

Real Estate Competition Fosters Resilience

In the case of coastal real estate, a state such as North Carolina competes with other coastal states such as Virginia. If North Carolina is not addressing sea level rise and Virginia has an edge in addressing the issue (either because of topography or due to active infrastructure investments), then Virginia has strong incentives to advertise its competitive advantage. By broadcasting these verifiable claims, Virginia's self-serving efforts further resilience by making North Carolina accountable for underinvesting and hiding the emerging risks. In this sense, spatial competition for jobs and people fosters resilience if the population is risk averse and aware that it is not fully informed about the emerging risks.

In December 2018, Virginia's governor Ralph Northam issued Executive Order 24, titled "Increasing Virginia's Resilience to Sea Level Rise and Natural Hazards." The document states: "This increase in extreme weather events and natural disasters will continue to have a profound impact on Virginia. It threatens public health and safety, our environment and natural resources, and the economic well-being of the Commonwealth, including our ports, military installations, transportation infrastructure, tourism assets, farms, and forests. We must act now to protect lives and property from multiple threats and reduce taxpayer exposure through fiscally responsible planning."[21]

The executive order creates a new state resilience officer position to coordinate the state's adaptation efforts. Short-run action items include a review of vulnerability of commonwealth-owned buildings, a unified sea level rise projection for commonwealth-owned buildings, and an investigation of the Virginia Coastal Resilience Master Plan: "The Commonwealth of Virginia has a responsibility to assist local governments in reducing flood risk through planning and implementing large scale flood protection and adaptation initiatives."

It is notable that the Virginia approach differs from North Carolina's efforts. According to a report in the *New York Times,* in North Carolina "a 2012 law, and subsequent actions by the state, effectively ordered state and local agencies that develop coastal policies to ignore scientific models showing an acceleration in the rise of sea levels."[22] Each investor in real estate will face a choice concerning whether North Carolina or Virginia is now a more attractive area for investment. Virginia's efforts also put more pressure on North Carolina's leaders to reconsider their past unwillingness to factor in sea level rise. Virginia's investments in upgrading disaster planning can benefit North Carolina if coastal officials from North Carolina visit Virginia's officials to learn best management practices. This possibility of imitation and adoption of good ideas dovetails with the economist Paul Romer's vision that codified plans can be scaled and used by multiple entities at the same time. In this sense, resilience planning ideas are public goods.

Mitigating Real Estate Risk

Climate economics research seeks to measure which cities will lose the most from experiencing warmer winters and warmer summers.[23] Such research begins with the point that real estate prices are high in temperate cities such as Seattle and San Francisco. These cities have warm winters and cool summers. Using cross-city real estate data, these researchers seek to estimate how much of the price premium for home prices in San Francisco relative to Chicago is due to San Francisco's better weather. Climate change, by raising San Francisco's summer heat, partially attenuates the unique San Francisco quality of life.

Such prospective studies are based on insights given existing technologies. The history of innovation teaches us that when there is a large market demanding new solutions, entrepreneurs and engineers work together to design such solutions. Their innovations help to offset the challenge. The open empirical question pertains to how effective these innovations are in fully offsetting the challenge and at what cost. As we have seen in the case of air conditioning penetration, more and more Americans now have access to such cooling technology and deaths from extreme heat have sharply declined in recent decades.

The rise of Las Vegas offers an interesting case study on the unexpected changes over time in how cities grow. Circa 1940, Las Vegas was home to roughly ten thousand people. Improvements in transportation and marketing, legalized gambling, warm winters, and increased access to air conditioning have all played a role in the city's boom. While its winters are warm, its summers are too hot to spend time outside. Because everyone knows this, there are ample indoor leisure options as entrepreneurs have created recreational opportunities (such as indoor gambling, indoor basketball, and indoor shopping malls).

Singapore is both highly productive and extremely hot. This city-state is known for its quality of life. Singapore's founding father, Lee Kuan Yew, was asked what the keys were to Singapore's fast economic growth over the preceding fifty years. He replied, "Air conditioning was a most important invention for us. . . . It changed the nature of civilization by making development possible in the tropics. Without air conditioning you can work only in the cool early-morning hours or at dusk."[24]

Economists have examined how the diffusion of air conditioning insulates the population and workers from extreme heat.[25] Air conditioning was called the "handmaiden of Southern growth" because it allowed households and firms to enjoy the warm winters in the South while suffering less from the heat and humidity of summer. In both the case of the American South and the nation of Singapore, real estate in these areas would be much less valuable today if air conditioning had not proved to be so effective and a relatively low-cost way to offset heat and humidity.

The risk that a piece of real estate faces is specific to the city and neighborhood where it is located. Consider coastal flood risk. The first determinant of flood risk is the topography of the property. Some properties are at greater risk of flooding than others. Those parcels of land that feature less risk because they are on higher ground will sell for a price premium. The second determinant of flood risk for a property depends on local investments in flood control. The locality can invest in many different costly strategies to protect the area from flooding. Such community investment benefits all property owners in the area. Local voters will support a public goods risk mitigation project such as a sea wall if the private benefits they will receive are greater than the extra taxes they will have to pay. Local property owners would prefer that the US Army Corps of Engineers implement and pay for such a project. The gains to the local community hinge on each household's risk aversion (more risk-averse households will be willing to pay more for government protection) and their perception of how effective the public investments will be in fending off sea level rise and flooding. Homeowners can also protect themselves by investing in precautionary activities such as elevating their home. There is an interplay between these public and private investment decisions. If a community believes that public investment is ineffective, it will invest in little public protection, and each household will invest in more private self-protection.

Gentrification and Climate Resilience

In July 2018, the *New York Times* published a piece discussing recent waterfront development in the New York City area. The piece pointed to a trend that the Sheepshead Bay area in Brooklyn was gentrifying and densifying as tall buildings were being built. Middle-class waterside communities featuring single-family homes were being knocked down.[26] If a neighborhood gentrifies due to a shift in which residents now live there, this changes the character of the community. The costs are borne by those incumbent middle-class households who are renters who cannot afford to buy or rent the new coastal housing. They lose their local place-based networks, and traditions disappear as longtime residents move away. The benefits of this change accrue to the incumbent homeowners who can sell their land for a large amount of money and to the new wealthy residents who move into the upscale buildings.

Tall, expensive new housing is being built on the land that was occupied by middle-class households. The new towers that are being built on the water will house the rich and at high density. The hundreds of millions of dollars invested in these projects create the capital and the necessity to bring in the best engineers to plan for risk contingencies. A key question pertaining to such expensive housing complexes is what stress tests do the builders place on such buildings to reduce the probability of catastrophic failure? Put simply, does engineering expertise and sufficient capital investment effectively counter future storms at the intensity level of Hurricane Sandy? Or are thousands of rich people setting themselves up for future disaster as they are lulled into a false sense of safety? We take gambles every day of our lives. The key economic issue here is whether the decision makers are aware of the gamble they are taking in this case.

As middle-class people sell their coastal homes that are then demolished to build new, safer apartment towers for the rich, a distributional question arises. On the one hand, this is a mutually beneficial trade such that the party who has a comparative advantage in coping with the new risk ends up controlling it. The middle class are engaging in a voluntary transaction where they are paid for their asset and their risk exposure declines as they move to another area. An ethicist might say that the win-win would be to permit the incumbent middle-class households to move into the higher-quality new structures. This would allow them to keep their networks and their old way of life in new buildings. The middle-class incumbents are unlikely to be rich enough to afford apartments in the new buildings. Should the nation's taxpayers subsidize their units? Do the incumbents have a property right to live in their original area?

In cases such that single-family homes are being demolished and converted into luxury apartment towers, then richer people will be moving close to the hazard (the rising sea). Such new apartment buyers may not be risk lovers. Instead, they may be risk-averse people who seek out the rare bundle of a new apartment with great views. Such condo buyers may be technological optimists who believe that the building has employed the best engineers to design the building properly to cope with risk. This raises the issue of how we as individuals trust the predictions of climate models. If the rich apartment buyers are overconfident with respect to the ability of engineers to fend off sea level rise, then they are at risk to be badly hurt by subsequent sea level rise. Given that these apartment buyers are paying millions of dollars for these assets, they will recognize that this investment will be worth much less in the future if the building is at risk from significant flooding. The prospect of losing much of their housing equity prompts these bidders for these units to do their homework before buying such a unit. This in turn creates a market for housing consultants with expertise in judging flood risk to offer such apartment buyers a second opinion before the investor actually purchases one of these units.

Who Chooses to Live in Harm's Way?

The population of Phoenix roughly doubled between 1980 and 2010.[27] Housing developers built hundreds of thousands of new homes. Developers will be willing to bear the costs of building such new place-based assets if they believe they can sell these properties. People will demand these assets if they believe that they can have a good life in terms of employment, quality of amenities, and raising their children in such a location. If extreme heat and drought risk associated with climate change threatens the quality of life in Phoenix, then developers would build less new housing there because they would anticipate that they could sell such new housing for less money and this would lower their profits. Recent research has documented that new housing supply construction is lower in counties that face a greater risk of sea level rise and where the local residents are more concerned about climate change.[28] If Phoenix is expected to be considerably warmer in the summer soon and if new potential residents cannot cope with such heat, then developers will build less housing.

In desert areas such as Las Vegas and Phoenix, there is extremely valuable housing even though these areas are hot and far from the coasts. A paradox arises. The popular media publish stories about the coming collapse of Phoenix due to extreme heat and drought at the same time that multimillion-dollar homes are built and sold.[29] A behavioral economist would posit that the new buyers are ignorant of the risks that their assets will face in the medium term. This theory predicts that such foolish buyers will later regret this investment.

An alternative explanation focuses on buyer diversity. Phoenix offers new, spacious homes that are much cheaper than their Southern Californian counterparts. The people who buy such homes are not a random set of people. People who love to jog outdoors during summer will not buy in Phoenix! They either live somewhere else in summer or do not mind not going outside for those months. The Internet provides ample lessons on how to adapt to the heat of Phoenix.[30] Economists often talk about comparative advantage. This term refers to what a person (think of LeBron James of the NBA) or a nation (think of France producing wine) is relatively good at producing. In a similar way, there is comparative advantage with respect to locational choice. People who enjoy staying inside have a comparative advantage in living in Phoenix.

A pessimist might posit that Phoenix will run out of water and thus will face increased drought risk. Economic logic suggests that water prices must reflect this scarcity, and then the laws of supply and demand will determine the allocation of this increasingly scarce resource. The cost of electricity and water is not high right now in Arizona. According to the Energy Information Administration, the average monthly electricity bill in Arizona in 2014 was $120.51. According to the city of Phoenix's finance department, the average single-family home in Phoenix has a monthly water bill of $37.75.[31]

The consumer city (shopping, culture, restaurants) is a major part of the modern urban economy.[32] If extreme heat on a given day discourages people from leaving home to shop and eat, then retailers have strong incentives to address this because they will lose revenue if people hunker down on such days. Singapore's upscale malls on Orchard Row provide one example of how the consumer city evolves in an extremely hot and humid place.

An extremely hot place will develop indoor culture and fun. The consumer city can prosper in the heat (think of the Middle East oil-rich nations). Yes, this air-conditioned Xanadu requires electricity, but this adaptation will be fueled by ever cheaper renewable power. As crime falls across the United States, people will be willing to spend more time out at night when it is cooler. In Singapore, the city comes to life late at night. Around the world, areas that are exposed to extreme heat reconfigure their day to reduce heat exposure. This lifestyle is not for everyone, but millions of people reveal a desire to live in Phoenix and Las Vegas despite the summer heat.

In my past work, I have documented that people are willing to pay a premium (and this premium has grown over time) to live in places with warm winters and cool summers.[33] California is one of the few places in the United States that offers this combination. Most places in the South have mild winters and hot, humid summers. The enjoyment we gain from indoor leisure during summer will play a key role in determining whether real estate in extremely hot places keeps its value.

Real estate economics research posits that the price of homes in risky areas may remain artificially high in the short run and medium term and then collapse when investors realize the actual high risks such properties face. At the heart of such models is the theory that investors will eventually realize that they had been underestimating the true and rising risk.[34] This fall-off-the-cliff pricing hypothesis implicitly assumes that few real estate buyers or incumbent homeowners are aware of the ambiguous risk. Suppose that incumbent homeowners are aware of the impending risk. They face a decision concerning whether they defer maintenance and invest less in the quality of the home. If they are sure that everyone will soon agree that such homes are at risk, then the resale value of the home will be low and they will be more likely to allow the home to depreciate. Alternatively, if they believe that there are overoptimists in the population or investors with behavioral biases, then they may invest in upkeep and seek to sell their property to one of these investors in the future. In this case, the optimists (and the climate deniers) will be more likely to purchase these assets and learn only when it is too late that the property is now very risky because of its location.

Real estate developers also have an incentive to engage in similar calculations in deciding whether to build new houses in increasingly risky places. If the developers believe that potential buyers are aware of the risk, then they will build fewer houses because the forward-looking buyers will bid less aggressively for them. If the developers believe that the buyers are myopic and underestimate future extreme heat or flood risk, then they will be more likely to build new housing in risky places and sell them to these naive individuals. While some investors have over-optimistic expectations, I do not believe that investors would make such a large investment in a down payment on a home without engaging in their due diligence. Just as people research the quality of the local schools and crime reports in a neighborhood they are considering, potential home buyers will increasingly seek out this climate risk information.

A person who buys a piece of real estate in a place facing rising climate risk may be an optimist about future technological advance such as infrastructure upgrades that reduce climate risk. The owner of the asset holds an option to invest in such future technologies. Incumbent homeowners have strong incentives to search for solutions using private goods and local public goods to offset the new risks posed by climate change. If they do not devise self-protection strategies, then their major asset that they rely on to fund their retirement (their home's value) will decline in value.

Resilient Real Estate

Homeowners have a growing set of strategies for protecting coastal real estate without retreating. Options include putting one's house on stilts.[35] Other strategies include smaller actions, such as installing foundation vents and sump pumps. Foundation vents, a form of wet floodproofing, allow floodwater to flow through your home rather than pool around it. Other advice includes applying coatings and sealants and grading one's lawn away from the house.[36]

At any time, the real estate capital stock is a mixture of mostly older structures and some newer structures. Given that building codes affect only new real estate structures, it can take decades for such codes to apply to a large share of the existing structures. Hurricane Sandy severely damaged or destroyed more than eight hundred buildings in New York City; of these, 95 percent were more than fifty years old and built before modern building codes. Mayor Michael Bloomberg has argued that improved building codes contribute to resilience because the buildings are of higher quality and sturdier.[37]

Enhancing building codes in flood and fire zones imposes costs on developers because these new safety features require new expenditures, but they do offer increasing expected benefits as new risks emerge. Risk-averse people will be willing to pay more for objectively safer housing in areas that now face more risk. For example, if new building codes require that special materials be used in fire zones that are more flame resistant, then this reduces risk. Building code enforcement means that all structures in a given area have been upgraded to reduce risk. Such clusters of safer structures offer further safety for each home. Imagine a case where one home is built with fire-resistant materials but all of its neighboring homes are not built to the same code. These neighboring homes' underinvestment in resilience poses a threat to the safe home if a fire breaks out. This logic highlights how enforced building codes reduce negative spillover externalities and avoid a domino effect in which fire spreads from an unprepared home to the neighboring homes.

As real estate buyers become more aware of the new risks that areas face, real estate developers are increasingly likely to build buildings with resilience features. These features are customized to address the challenge that the climate science states the area will face. For example, in Florida parcels face extreme hurricane risk. Developers have responded by building structures such as 1450 Brickell in Miami. This is a thirty-five-story Class A office tower built with impact-resistant and hardened materials to withstand flying debris and winds up to three hundred miles per hour. It remains to be seen whether this building is as sturdy as the boosters claim. Future natural disasters will offer a laboratory experiment that will allow us to learn how durable these new buildings are. If buildings such as 1450 Brickell perform better during a natural disaster than nearby older buildings, then this natural experiment will reveal the value of climate resilience infrastructure in reducing risk exposure.

One example of how developers are achieving resilience by partnering with risk assessment firms is the work of the Frisbie Group in Palm Beach, Florida. This firm is building high-end retail and restaurants and real estate while recognizing that such coastal assets will face sea level rise risk. The managing director of this firm states: "Being on the water isn't necessarily a risk if you are building to the highest standards. It's when you are not building to the highest standards and the right elevations that there is a risk."[38]

In my past research, I have documented that solar homes and other green real estate sell for a price premium.[39] If recent home buyers are sufficiently risk averse, then we will see a similar price premium for climate-resilient real estate. Such buyers and renters would need to be aware of the risk and to fear the risk. In this case, they would be willing to pay a premium to live in a safer structure in an increasingly risky place. On the supply side, the cost of building such resilient buildings could fall over time.

The *New York Times* reported in 2018 that "communities in the West are implementing their own strategies. In San Diego, new subdivisions are being built with fire-resistant designs and materials so residents can stay safe in their homes while the fire burns around them, instead of risking evacuation and the perils of clogged roads."[40] As more people seek out safer housing, this rising demand creates greater incentives for firms that can create risk-resilient products. In fire zones, construction that is fire resistant will be increasingly valued. The induced innovation hypothesis posits that such products will become cheaper over time due to scale economies and specialization and competition to produce them.

Improvements in information technology facilitate increased real estate resilience. Consider Nantum software.[41] This software allows owners of multiple real estate buildings to use a dashboard to observe the performance of the various buildings in real time. To quote the company's webpage, "Nantum's® algorithms continuously improve over time, increasing building efficiency, energy reduction and cost savings, while enhancing tenant comfort and retention." By relying on such software, a real estate manager can quickly diagnose new challenges and devise optimized solutions to minimize the impact of a heat wave or a pollution outbreak on the building's occupants.

This example highlights the synergies between human capital and the scale of activity. Owners of a single building may not have the scale of operations to make it profitable for them to invest their time and money to purchase access to this software and to become experts in using it. In contrast, owners of several buildings have the scale to make such an upfront investment profitable and to hire a high-quality engineer to oversee the optimization of all the buildings. In this sense, the sheer scale of a larger real estate portfolio creates an incentive to seek out using big data and to use a computer interface that enables a trained engineer to use this information to fine-tune the buildings to maximize their performance even under extreme conditions.

An alternative strategy to building and operating resilient real estate is to build either modular real estate that can be disassembled if risks such as sea level rise turn out to be severe or temporary nondurable structures such that owners can walk away from these at relatively low loss if the land is flooded. Those who live in climate-risky areas can build less durable structures. In a 2017 paper, I argued that less durable capital offers its owner an option to wait and see.[42] If the new risks are as bad as the climate scientists claim, then the real estate owner can earn a rental flow on the short-lived asset and then walk away. If the asset turns out to have solid fundamentals because the climate scientists exaggerated the risk, then the asset owner can make a larger investment in a more durable, long-lived asset. An even more extreme version of this logic would be for coastal real estate investors to build modular housing that can be disassembled and the capital salvaged if sea level rise accelerates.

Grappling with Managed Retreat

There are many emergent strategies for dealing with different climate change–posed real estate risks. Despite this adaptation progress, sea level rise will cause some areas to grapple with managed retreat. Some urban planners advocate a planned retreat such that housing along coastal areas is converted into green space and wetlands that provide flood protection to the rest of the area. Ongoing research has investigated the cost effectiveness of such land transitions as private property is set aside for public space that offers resilience services.[43]

In a managed retreat, property owners are asked to vacate properties. These property owners will suffer a double loss. First, they will be displaced from their homes and their way of life where they have established roots. Second, they will suffer an asset loss because they will not be able to sell their homes to another private buyer. Many of these property owners seek to sell their properties to the state at a high price. The state wants to purchase adjacent properties so that the land can be reconfigured as wetlands to reduce flood risk for inland properties. An ethical issue will arise pertaining to what is fair compensation for a property that faces increased flooding. One study in real estate economics documents the reluctance of homeowners to sell their homes for a price less than the original purchase price.[44] If the incumbent owners refuse to sell for a low price, then the state may face the issue of using its eminent domain power to clear out property owners. If these areas do not rezone for higher-density buildings, such areas will face additional challenges of perhaps having too few children for the local schools and too few taxpayers to pay for basic services. Such shrinking jurisdictions will be likely to seek to merge into the county's jurisdiction to provide basic resources. In this sense, the climate change costs imposed on one jurisdiction can affect the finances of that jurisdiction's county (and hence the rest of the taxpayers) if the jurisdiction's tax base shrinks.

7

The Market for Big Data Facilitates Adaptation

Back in the 1970s, the Internet and the cell phone as we know it did not exist. Real-time information about new challenges could be learned only from watching television or listening to the radio or reading the newspaper. Individuals could not widely share their new information about an emerging challenge such as a nearby smoke plume. In those days, a family who needed a new air conditioner would call the local hardware store to see if it had one in inventory. If the store was out of stock, you could wait weeks for the new inventory to be delivered. Retailers did not have computers to help them to identify inventory gaps. Instead, long lines of complaining customers nudged store managers to make phone calls to track down possible suppliers of such products. These lags created delays in purchasing such adaptation-friendly products.

Today, our constant access to the Internet reduces the likelihood that we are unaware of new threats. Cell phone messages, Facebook, Twitter, and Google provide constant sources of new information for the curious. Each of us will be better able to adapt to climate change if we recognize that it poses many known unknowns. Learning facilitated by the Internet means that unknown unknowns become known unknowns.

In this chapter I explore how the rise of information technology makes us better able to adapt to new climate risks. Every sector of the economy, including households, firms, and governments, is both supplying more real-time data and demanding more data. More and more young people are training to become data scientists to crunch these data. This information processing helps us to build up our individual and collective resilience to risk.

The Demand for Real-Time Data

In the age before the Internet, cell phones, apps, and text messages, you could have no idea that a major heat wave was on the horizon or that a fire was burning nearby. Back then, one could be caught by surprise when a shock occurred. The more time you have to prepare for a shock, the better you'll be able to cope through strategies ranging from evacuating an area or heading to higher ground or going to a cooling center.

More and more data are being collected across the world and over time monitoring temperature, pollution, fires, and flood risk. In real time, there are apps telling us about our localities' heat, pollution, and natural disaster risk. Such apps can report this information because of the ongoing decrease in satellite technology and drone costs. As this sensor technology improves in quality and declines in price, it becomes increasingly possible and affordable to have real-time coverage of emerging risks.

Given the rising fire risk in California, there is an increase in the demand for app technology that shares real-time information on elevated local air pollution levels caused by nearby fires. One study examines how people use a new air pollution app with several distinctive features. This app informs users about local air pollution levels and asks them to provide current data about their self-assessed health. By collecting these data at multiple points in time, this app teaches its users about their own susceptibility to pollution. The app also shares information with each individual on how the pollution is affecting their neighbors' health. Such peer comparisons can also provide useful information.[1] This app accelerates pollution adaptation because it creates two-sided learning. The users each learn about their own susceptibility to pollution, and the researchers collect new data to study how the aggregate user base is affected by the pollution. For example, if the app is used by fifteen thousand people, then the research team that observes all of the data can measure the impact of pollution on health for each person and thus have a much better sense of pinpointing who are the most susceptible people and who are the least pollution susceptible in the user population. By capturing this population diversity, public health authorities can better target interventions to protect those at the greatest risk.

Increased access to climate big data will help people to update their beliefs about new risks. Some people underestimate future risks as they form their expectations of the future based on their recent experiences. In an area experiencing climate change, this backward-looking approach will underestimate new risks. In September 2019, the *New York Times* reported than many homes in North Carolina face flood risk but sit outside the boundaries of the current hundred-year floodplain. The *Times* argued that this set of homeowners is less likely to buy home insurance.[2] If these homeowners could be nudged to purchase a flood-risk report from one of the new climate risk report sellers, they might update their beliefs about emerging risks and then be less at risk from these anticipated future events.

When my family searched for a new home back in 2008, the multiple listing service provided information on each home's price, square footage, and other physical attributes. Zillow and other real estate websites provide information on local neighborhood features such as public school quality, crime levels, and neighborhood walkability. In the near future, more data such as a neighborhood's average annual temperature by month, the property's recent exposure to past flooding, and whether the property is located in a fire zone could also be reported.

On Twitter and Facebook, people can tap into many sources of information. Some people trust their friends' opinions more than they trust the news or government's announcements. The Internet provides easy access to many perspectives about an emerging challenge. While people may want to hear information that confirms their initial views, it is common sense that people also want to be informed about emerging risks.

Big Data in Daily Life

New real-time technology daily monitors an individual's well-being. With the Fitbit watch, you can monitor your temperature, pulse, and blood pressure to see how you are performing on an extremely hot or polluted day. These insights help you make better decisions that you will be less likely to regret. During a natural disaster, a Twitter feed from a trusted source helps people on the move to have a better sense of which areas are safe. This information facilitates an orderly evacuation and thus cause less dislocation and panic.

The big data revolution provides each of us with clues about our health and well-being as we can monitor our sleep and pulse using an app. On extremely hot and polluted days, people can use their smart phones to learn about how these climate shocks affect their daily quality of life. People can measure how sensitive they are to heat and pollution and then change their plans for that day accordingly. I published a coauthored paper documenting that urbanites in China are more irritable and moody on extremely hot days and polluted days. We were able to document this because we had access to hundreds of millions of individuals' social media posts and we knew the day and location where each post was generated.[3] If people are aware of their own predisposition to be angrier on such days, they can counter the heat through such measures such as meditation or taking a siesta to cool down. Such investments foster self-control and reduce the likelihood of making a mistake by losing one's cool at work or by triggering a violent encounter with a stranger. Sleep patterns can be disrupted by high heat; new apps and devices allow one to monitor one's sleeping patterns and to make adjustments accordingly.[4] These quantitative insights fuel behavioral change and a demand for new products to facilitate adaptation. If people are aware that they are underperforming and stressed out because of the heat, they will seek solutions and this creates a new market for products that are effective.

In Beijing on a cold winter morning, people check their smart phone app to learn how polluted the morning air will be. China's Ministry of Environmental Protection issues pollution alerts. In my recent work, we document that people purchase more masks and air filters on days when the authorities announce that it is very polluted.[5]

In the social media age, people have access to many sources of information. Some people rely mainly on their friends for information. In this case, a platform like Facebook will play a key role in helping individuals to cope with emerging challenges. If people mainly trust their friends for advice, then they can tap into their Facebook networks for trusted information. In the case of combating tuberculosis in the developing world, peer referrals have helped to target those who are vulnerable and need encouragement to be tested and to seek out treatment.[6]

As the cost of information access has plummeted, the supply of information has exploded, and many are concerned that the quality of information has declined as the supply of fake news has increased. Platforms such as Facebook have wrestled with the challenge of whether it is their job to weed out false information. At a time of political polarization, climate change has been labeled a progressive issue. I continue to believe that political conservatives will quietly research the specific climate challenges their real estate and businesses now face, but this claim merits testing.

Big Data Facilitates Introducing Dynamic Pricing for Electricity and Water

Every household uses electricity and water to produce basic goods such as being clean, storing food, reading, and educating one's children. Since water and electricity are essential for households to thrive, officials have been cautious about exposing households to higher prices for these goods. With the installation of smart meters in people's homes and in commercial real estate, one's water and electricity consumption can be measured every fifteen minutes. In the past, people would receive such a utility bill once a month and would thus have no idea what their total consumption was in the middle of the billing cycle. Access to real-time information on consumption opens up the possibility of introducing time-varying prices for water and electricity. Such dynamic incentives would sharply reduce the likelihood of drought and power blackouts even during times of limited rainfall and high electricity demand.

As climate change increases drought risk and raises summer electricity demand, water prices and electricity prices need to rise at such times to signal scarcity. Dynamic pricing for water will mean that during times of drought, the price of a gallon of water increases. During peak times, such as the late afternoon on a summer day, the price of electricity will rise. This introduction of dynamic pricing could be bundled with lower prices off-peak so that consumers could actually face reductions in average prices. Poor people could be protected against such price spikes by giving this group an income transfer to offset the higher prices they will face at peak times.

Higher prices for electricity would induce consumers to change their consumption habits. People would turn off their computers and televisions in the late afternoon in order to consume less electricity. In aggregate, this behavioral change would protect the grid against blackout risk because demand and supply would be in balance. There is always a consumer (whether residential, industrial, or commercial) who is willing to reduce electricity consumption if the price she or he pays goes up.

Researchers have used electric utility data at the residential customer level to document that households respond to critical peak pricing by sharply reducing their electricity consumption.[7] Natural experiments conducted in Washington, DC, enrolled households in a pricing program such that those who agreed to participate anticipated that on certain days (perhaps twenty a year) the electric utility would raise hourly electricity prices and alert these households to the exact timing when these critical peak pricing events would occur. The researchers used the utility's administrative data on hourly electricity consumption by customer account to study consumption dynamics. During times when electricity prices were higher, households responded by consuming less. Such research does not tell us how much comfort the household lost during those hours. If the thermostat is raised from 72 degrees to 74 degrees (in order to reduce electricity consumption), how much discomfort did the household experience?

Recall that during the global energy crisis of the 1970s, President Jimmy Carter gave a major speech on television while wearing a sweater. He was trying to nudge Americans to put on more clothes during winter rather than cranking up the heat (which uses heating oil). President Carter was quoted as saying back in February 1977:

> We must face the fact that the energy shortage is permanent.
> There is no way we can solve it quickly. But if we all cooperate and
> make modest sacrifices, if we learn to live thriftily and remember
> the importance of helping our neighbors, then we can find ways to
> adjust, and to make our society more efficient and our own lives
> more enjoyable and productive. Utility companies must promote
> conservation and not consumption. Oil and natural gas companies
> must be honest with all of us about their reserves and profits. We
> will find out the difference between real shortages and artificial
> ones. We will ask private companies to sacrifice, just as private
> citizens must do.

> All of us must learn to waste less energy. Simply by keeping our
> thermostats, for instance, at 65 degrees in the daytime and
> 55 degrees at night we could save half the current shortage of
> natural gas.[8]

In 1977 Carter did not anticipate the big data revolution. Today, we can achieve his goal without nudging the public to engage in voluntary sacrifice. Instead, the adoption of market incentives can bring about the same outcome.

In an age of increased drought risk and sharp increases in electricity demand during the peak summer season, the big data revolution allows for real-time dynamic pricing such that prices rise for water and electricity during times of high demand and low supply. By exposing people to scarcity signals, this creates incentives for demanders to economize on scarce resources, and this triggers induced innovation because companies now have a profit incentive to design more efficient products.

Using their smart phones, consumers of power and water can stay up to date in real time about the prices of such goods. By raising prices during peak demand times, the local electric and water utilities signal to all consumers that they have an incentive to conserve their consumption of an increasingly scarce resource. Some will set the thermostat higher while others will watch fewer movies during the afternoon heat. Any time a price rises, our real incomes are lower. To protect the poor, an income rebate could be given to households below the poverty line. Such an income transfer to the disadvantaged would mean that the price spikes achieve the economists' goal of signaling scarcity at peak demand times but the poor will not suffer a loss in their purchasing power (because they will receive an income transfer).

If more people and firms are willing to sign up and participate in dynamic pricing plans (perhaps because they receive an incentive to do so), then this will create a new market for more energy-efficient appliances.[9] If an existing air conditioner uses a given level of power on a 95-degree day and the price per unit of power doubles, the operating cost of this device has gone up. Its owner will be willing to pay more for an air conditioner that is more energy efficient. If there are enough demanders seeking such green durables (because more people are enrolling in dynamic pricing), then this will induce entrepreneurs to design these products. In this sense, demand creates supply.

Consider the example of farmers' water consumption. Farmers consume much more water than urbanites. For historical reasons, many farmers face extremely low water prices. As some areas have raised prices for farmers' water, farmers now have an incentive to spot inefficiencies on their land. They have been hiring drones to fly and look for irrigation leaks. This creates a new market in high-resolution drone technology in order to economize on more expensive water. If the price of water had remained artificially low, then this green technology would not been developed and deployed. Resource pricing encourages adaptation and less waste.

The Role of Big Data in Fueling New Climate-Resilient Insurance

In recent years, the health insurance industry has increasingly relied on data analysis to offer a menu of health insurance policies tailored to the needs of the diverse population. Young people have different health insurance needs than older people. People with preexisting medical conditions and people who are highly risk averse seek different health insurance coverage. A young, healthy person who does not mind some risk might seek out an insurance plan that mainly covers catastrophic medical events. Such a young person would pay a premium for a plan that covers such an event but would face a deductible for medical office visits. Using big data on millions of other young people with similar life histories, the insurer would crunch these data to calculate the probability that this young person will experience a catastrophic medical event in the next year. If this probability is low, then the insurer can still earn a profit even by charging a relatively low insurance premium.

In this age of big climate data, the same logic applies to insurance pricing. In the case of real estate, the major climate risks include flooding and fires and other natural disaster risks. In the face of these new climate risks, new possibilities of private-public partnerships will arise. The National Oceanic and Atmospheric Administration operates many satellites that provide crucial remote-sensing data about emerging new risks. Private sector firms use this information to create easily grasped information summaries about the specific challenges that different parcels of land face focused on flood risk, extreme heat exposure, and natural disaster risk exposure.[10] Examples of these firms include Four Twenty Seven, Jupiter, First Street Foundation, and Coastal Risk Consulting. Firms like these are using government-collected data and partnering with academics to create a user-friendly interface to provide geographically refined information about the emerging challenges that specific parcels of land now face and are likely to face in the medium term. In this sense climate big data is an emerging growth field.

To enhance my own research, I have formed a partnership with the First Street Foundation. Under our agreement, I receive no financial compensation or monetary research support. Instead, I have access to some of the firm's spatial data, which allows me to conduct new research. In one ongoing project, we are exploring how the city of Baltimore is likely to be affected by sea level rise. This project will inform the city's planning decisions over infrastructure investment and land zoning.

The demand for high-quality climate risk probability models increases the likelihood that entrepreneurs will enter this space and create more accurate risk models. In California, it has been estimated that a million homes are located in fire zones. If each of these homeowners demands insurance, then this creates a large market for insurance firms to invest in the modeling capacity to accurately predict spatial fire risk. Such risk models would be used by insurers to set insurance prices, and then homeowners could choose whether to buy such a policy. Risk-averse homeowners would be more likely to buy the policy. Given the fixed cost of doing the basic research to create a high-quality fire probability statistical model, for-profit firms will incur this upfront irreversible cost only if they expect that a large number of people will demand the predictions based on the model. In this sense, expected misery from climate change induces a new solution (improved fire forecasting) that creates the new market for fire insurance. An ugly scenario could arise if the actuarial fire risk is rising but no fire modeling entities are conducting the research to pinpoint which properties face which new risks because these entities do not believe there is sufficient demand to allow them to earn a profit on this investment. In this case, the National Science Foundation (NSF) could fund such basic research.

Although I am highly optimistic about the creation of these new markets, it is important to note that the federal government can slow this private sector initiative. The government's active role in subsidizing insurance in flood areas reduces the aggregate demand for private insurance, and this reduces the profitability for private insurance firms from investing in pinpoint spatial knowledge. In this sense, the well-intentioned federal policy to help those who suffer in disasters leads the private sector to retreat from this field. This is socially costly in terms of slowing the rise of resilience if the private sector has an edge in researching emerging risks and incentivizing households to move to safer places and to build with better materials.

The New Economy Firm's Role in Fueling Adaptation

Each of the major Silicon Valley companies plays a role in helping us to adapt to new risks. Consider Google. The ability to search for any piece of information and for this information to be delivered in seconds helps an individual make better choices. For example, if a person in Los Angeles does not know where the nearest public cooling center is, she or he can quickly find this information and then use Google Maps to find out the shortest commute time to get there.

Google instantly allows individuals to search for low prices for products they need to facilitate their adaptation. Examples range from backup power generators to new air conditioners. Facebook plays a key role by allowing people to learn from their social network and to signal that one needs advice from one's social network. Some people may be more willing to listen to their friends' advice rather than trusting information they read online.

During a crisis there is often a spike in demand for transport to a safer place, as well as access to shelter, food, and water. How such scarce resources are allocated will be affected by the sharing economy. New economy firms such as Uber will play a key role in helping us to build resilience by efficiently using our transportation capital stock to move people away from the danger. Most American households own at least one car, and this vehicle is parked most of the day. The brilliance of the sharing economy's core business model is to raise the deployment rate of such capital.

During Hurricane Harvey, numerous private canoes were used to rescue people. It does not require much imagination to anticipate that soon Uber and the other ride-sharing services will offer emergency relief during disasters. Poor people will be less able to afford these prices, but private citizens could pay for their rides as a form of charity. The ability of the private sector to augment the federal government's emergency first responders will help to save lives. Such competition in providing emergency services will improve response times.

Uber is a two-sided platform that connects buyers and sellers of transport services. Such services play a key role during a crisis when people need to move away from an affected area. For those with limited resources, there could be general PayPal accounts such that charitable individuals could donate funds during emergencies so that poorer people could pay for Uber services.

Uber is just one example of the sharing economy. During a crisis, Airbnb and Facebook will play a similar role as displaced households look for short-term housing. The Internet technology connects strangers, allowing them to trade with each other. Rather than relying on charity to take people in, the Uber market approach offers such households a financial incentive to allow strangers to stay with them. This supply-side incentive will lead to more aggregate disaster relief and a smoother adjustment process as those in need are directed by the app to locations that welcome them to stay there. The combination of charity and government emergency relief and the private sector raises the likelihood that more victims recover faster from future natural disasters.

In the aftermath of Hurricane Katrina, thousands of people were sleeping in the Superdome. In 2005, Uber and Airbnb did not exist. Would the Katrina-induced short-run housing, food, and comfort crisis have been less severe had these platforms existed? These platforms facilitate trade. During a crisis, there are people who seek transport and housing services, and there are other people who are willing to supply these services. These Internet platforms act as a central clearinghouse, bringing both sides of the market together to trade with each other.

Once people are rescued, the big data Internet firms help people who need shelter find a place to live through platforms such as Airbnb. Major retailers are stocked with supplies that help people cope with disasters. Economists have studied how household access to food is affected when Walmart opens up Supercenters. These centers feature low food prices, so fewer people who live in the vicinity of a new superstore report facing food insecurity challenges.[11]

Cities such as Los Angeles have been signing contracts with ride-sharing firms to collaborate in providing transit services. On extremely hot days, ride-sharing services can play a role moving people from parts of the city featuring little residential air conditioning to cooling centers. This reduces the poor's heat exposure.

The ability of firms such as Walmart to use its big data to see what products are in demand means that poor people, who in the past would have been completely reliant on governmental aid, will be at lower risk of suffering in a crisis because both such nearby superstores and the federal government (the private and the public sectors) offer aid.[12]

During a crisis Walmart will be unlikely to engage in price gouging. Large firms such as Walmart care about their reputation, and the goodwill that such a firm would receive for helping the less fortunate during a crisis would likely outweigh any short-run profit gains from charging $10 for a can of tuna. Uber has faced a backlash when it has enacted extreme surge pricing at times of high demand, such as New Year's Eve.

Corporate Big Data Analysis Facilitates Adaptation

Amazon, Facebook, and other major technology firms all rely on machine learning techniques and their access to enormous microdata sets to figure out how to target different consumer groups to raise their profits. As these firms seek to earn higher profits, this actually helps consumers to build up their resilience to new climate risks. As people search for products on Amazon's website, Amazon tracks what people search for and what they purchase when Amazon quotes them a specific price for a product. Because Amazon knows each customer's mailing address, Amazon can merge data on the local temperature and pollution conditions at a given place and time. With the customer's mailing address, Amazon knows a person's neighborhood and can use census data to merge such neighborhood demographics as average educational attainment, age, and race.

During a hot summer, Amazon can use its database to identify the subset of people who are likely to purchase an air conditioner if they are offered a price discount. For this group of price-sensitive consumers, Amazon will consider offering them a price discount through a coupon or a special sale. If Amazon lowers the price of the air conditioner for the price-sensitive consumers, then this will raise the probability that they purchase the product, and this accelerates the spread of air conditioning.

Amazon's role in facilitating adaptation can be seen by revisiting the severe heat wave that struck Moscow in 2010. Thousands of people in this cold-climate city died as this unexpected heat wave struck. The demand for air conditioners soared, but stores quickly sold out. In 2012, Hurricane Sandy knocked out power for thousands, and the demand for backup power generation soared. Fast-forward to today and imagine if the climate scientists generate short-run prediction models that more accurately forecast extreme heat or storms ten days ahead. Companies such as Amazon earn a profit from selling durable goods. Amazon already has a credit card on file for each customer, has created one-click purchase, and often offers next-day delivery. By leveraging the behavioral economics insight that consumers want to avoid hassles and desire immediate gratification (next-day delivery!), Amazon is positioned to earn large profits when disaster season arises or when an event is anticipated. In this sense, Amazon's eagerness to make more profits dovetails with society's goal of helping people be more resilient in the face of rising climate risk.

Given your past Amazon purchases, Amazon's use of artificial intelligence allows it to predict your needs as a function of your purchases and the geographic challenges you now face. Such a sophisticated company can target you with incentives (if it believes that you are price sensitive) to purchase goods that will help you play defense. The adaptation products available on Amazon range widely from air pollution filters to cooling suits. Consider just one example: for $99 (or $199, if you'd like to upgrade the quality) and free shipping, you can buy a three-day survival kit for two on Amazon.[13] The firm's vast number of customers creates a unique database that allows Amazon to foresee emerging buyer demand trends. If consumers in Portland begin to buy bottled water, Amazon will see this pattern and will restock water to sell there if its artificial intelligence models predict that demand will rise. These demand-side models allow Amazon to avoid inventory depletions, and this means that consumers will be able to buy the goods they need when they need them.

In our new economy, an inventory collapse such that no major companies are ready to meet demand could emerge only if each of these major companies relies exclusively on the same incorrect statistical model. Each company has its own inventory-stocking strategy based on its big data research. Such companies anticipate when hurricanes and heat waves may take place in different locations. Inventories will be cheaper to hold if land for warehouses near the cities can be purchased. If drones can fly at 150 miles per hour and deliver packages from distant warehouses to warehouses closer to cities (where the packages are then loaded onto trucks), then the costs of durable inventories fall even further.

Even Amazon does not sell everything. In a 2018 article probing the world's vulnerability to a pandemic, infectious disease expert Dr. Michael Osterholm was asked if there was a top myth he would like to dispel. He replied:

> I think people believe that because you can go on the internet and order something from Amazon and it's here tomorrow, that anything we need in the medical care field will be available in equal speed. We don't have stockpiles of anything beyond a limited supply the US government has of some medical products, which would be quickly exhausted if we are in a real pandemic. We have to anticipate these things, and we have to have plans. Right now, anticipation is the word that probably applies to the next 12 hours. What we need to understand is that it has to apply to the next 10 to 15 years.[14]

As demonstrated by the shortage of masks, ventilators, and other emergency supplies during the 2020 covid-19 pandemic, Dr. Osterholm was correct. A firm such as Amazon anticipates that such key medical supplies will be in extremely high demand only if an epidemic occurs. During such a time, the American people would revolt against being price gouged by Amazon. Amazon anticipates this and chooses not to stock up on this good.

The major firms are continually updating their models to reflect emerging demand for their products. When past correlations no longer explain the current data, a for-profit firm will ask why, because it is losing profit if it continues to rely on the outdated model. Such firms have a strong profit motive to update their predictive demand models and to use their big data to predict how price sensitive consumers are.

Firms that prove to be good at recognizing new climate patterns will earn higher profits, and those that are not nimble will lose profit. This evolutionary process facilitates corporate adaptation. Since customers purchase these products, they gain resilience benefits from the increased sophistication of firms to foresee emerging climate patterns.

With each visit to its website, Amazon collects more customer data. This allows it to estimate individual-level demand curves for every product and to measure local aggregate demand. For example, it can predict its aggregate demand for air conditioners as a function of their prices in New York City in a future summer. This information allows Amazon to choose a price for selling this product that maximizes profits. If the price is set too high, then Amazon will lose air conditioner sales. If the price is set too low, then the firm sells more air conditioners but loses revenue it could have earned from those households who would be willing to pay more for the product. Its detailed database allows it to identify those customers who are and are not price sensitive. It can then offer a price discount to the price-sensitive customers. As Amazon collects more data on individuals, it can pinpoint which customers are quite price responsive. By offering such individuals price discounts through coupons, Amazon sells more self-protection goods. If such a firm did not know you, it would charge all people the same price and would sell fewer of the goods.

Amazon's pursuit of profit will encourage it to pursue marketing strategies both before and after climate shocks. If climate scientists are able to anticipate a heat wave ten days in advance, then this gives Amazon the time to use its platform to engage in nudging and to advertise air conditioning and other cooling products to sell them before the heat wave begins.

In this case, a researcher who uses past data to predict climate change damages would overstate the costs of new heat waves and smoggy days. As Amazon's sales of climate change adaptation-friendly products increases, in part because of its successful marketing efforts, more people will have the goods they need to cope with new risks.

Back in the 1950s, the Harvard economist John Kenneth Galbraith argued that Madison Avenue advertising manipulated people by stimulating demand for products such as cigarettes that people did not inherently want.[15] An alternative view posits that Amazon recognizes that each person seeks to be healthy and comfortable but that some people have trouble identifying the most cost-effective products to help them to achieve this goal. In this case, such individuals will underinvest in self-protection if Amazon does not use its enormous data and machine-learning techniques to identify this set of customers. Due to the profit motive, Amazon has strong incentives to nudge such individuals to purchase products to help them be more resilient in the face of air pollution, heat, and other climate change–induced threats. In this case, Amazon's profit motive aligns its objective with society's goal of creating more resilience among those who might otherwise underinvest in self-protection.

Charging diverse people different prices for the same product raises important economics and fairness issues. Big data analysis allows firms to identify consumers who greatly value a product versus those who are more price sensitive. If such firms did not use big data, they would not be able to identify who is a price-sensitive customer and who loves a product so much that they would pay more to buy it. A for-profit firm has a profit incentive to treat these two groups of customers differently. It will charge the price-sensitive customers less than the price-insensitive customers. This price discrimination benefits those who are price sensitive.

Consider another example. The neighborhoods of Lents and Powellhurst-Gilbert in Portland, Oregon, often suffer flooding from Johnson Creek.[16] If property insurers cannot distinguish between more elevated homes (that face less flood risk) from less elevated homes (that face more flood risk), then all of these local homeowners will be charged the same price for flood insurance. Big data access such as access to Google Earth allows insurers to identify homes at less risk (those that are more elevated), and their owners will be charged less for flood insurance. This example highlights how big data is used to partition groups of consumers into different types (price sensitive, less price sensitive, or high risk/low risk), and these subgroups of consumers then are offered deals that depend on this refined information. The firms selling products such as air conditioners or insurance benefit from being able to use big data to refine their marketing strategy, but some subgroups of people (in this case, price-insensitive consumers or those whose homes are not elevated) do not receive lower price offers.

Synergies between Data and Computing Power

Jeff Bezos, the CEO of Amazon, has recognized that other firms do not have the analytical abilities of his firm. Amazon makes profits renting out such services to these firms. As he stated in 2017:

The most exciting thing that I think we're working on in machine learning, is that we are determined, through Amazon Web Services—where we have all these customers who are corporations and software developers—to make these advanced techniques accessible to every organization, even if they don't have the current class of expertise that's required. Right now, deploying these techniques for your particular institution's problems is difficult. It takes a lot of expertise, and so you have to go compete for the very best PhDs in machine learning and it's difficult for a lot of organizations to win those competitions.[17]

There are economies of scale in computing power, and Amazon is willing to rent out its services and its computer codes to help other firms solve their problems. Smaller firms increase their productivity and their ability to engage in advanced problem solving by using Amazon's infrastructure. It would be costly for smaller firms to bear a fixed cost of hiring the right people and then designing their own software. Amazon recognizes this point and rents out its services to these firms, and this enhances the productivity of these smaller firms. These services help smaller firms to see data patterns sooner and to appreciate how a diverse customer group is responding in real time to new challenges it faces.

The Federal Government Confronts the Resilience Challenge

The US government is composed of dozens of different agencies and functions. Climate change will pose distinct challenges for each of these entities, ranging from the Department of Defense (defending our coastal military bases in places such as Norfolk, Virginia) to the Department of Housing and Urban Development (providing public housing for the poor).

Big data plays an essential role here by allowing the federal agencies to diagnose the challenges they face and by creating a benchmarking metric for measuring whether different agencies are becoming better at adapting over time. Consider an example concerning the Defense Department: suppose it is found in the year 2025 that soldier productivity is higher on extremely hot days than it was back in 2018. This would support the claim that the Defense Department has made progress in adapting to high heat. Sensors in deployed army soldiers' uniforms could create data on the degree of heat stress these soldiers face while deployed. In the case of the Department of Housing and Urban Development, updating flood maps will help the department to understand emerging infrastructure risks its public housing now faces. By identifying emerging problems, the federal government will have the time to engage in preventive maintenance. These actions fuel adaptation.

If each agency in the federal government employs a chief resilience officer, these individuals will demand high-quality real-time information identifying new threats their agency faces. If the government can commit to clear standards for how it will judge who has the best model, then private sector firms will invest the capital and their employees will develop the human capital for studying the resilience problem. In this sense, the federal government can play a key role in directing applied scientists to work on an issue. The National Science Foundation achieves this by having targeted calls for proposals that reward scientists for making progress on a specific problem. The US government is large enough to have this market power to direct technological innovation. If climate resilience becomes a priority, then the various agency resilience officers can pool their resources to entice more entrepreneurs to compete. This is a different model of intellectual property creation than the Manhattan Project during World War II. That nuclear bomb creation project took place in private at a secret lab.

Big Data Creates an Adaptation Laboratory

The rise of big data has created a new market for data scientists. These computer programmers have the skills to work with huge databases and to manipulate these data using statistical techniques to discover new facts about our evolving economy. Major universities are scrambling to create big data majors, and statistics departments have been revitalized. At the University of Southern California, the engineering school offers a master's degree in data science that includes courses in the foundations of artificial intelligence and in probabilistic reasoning.[18]

Such skills can be applied to gleaning new insights from enormous data sets. Consider all the 911 calls across Los Angeles County over the past decade. Such a data set contains millions of records. A data scientist can map where these calls originate from and make separate maps based on 911 call categories such as reporting a crime or reporting the need for emergency medical care. These maps display patterns of the hot spots, identifying the geographic areas featuring the most 911 calls. Such information is crucial for alerting social workers and helping the mayor's staff to target resources.

In Chicago, the city government has been working with engineers and data scientists to install five hundred information-gathering nodes throughout the city.[19] By measuring data on air quality, climate, and traffic, this sensor network provides useful real-time information. The city posts the data in an open data format to allow concerned citizens, the media, and researchers to have real-time access to the data. Such information holds urban leaders accountable if there emerges clear evidence that certain neighborhoods and groups of citizens are suffering significant reductions in their local quality of life.

The creation of such a real-time database is a necessary but not a sufficient condition for bringing about political accountability. If nobody analyzes the data or if it is in a format such that the data cannot be compared across cities at a point in time or for a given city over time, then such data will not provide a benchmark for quality of life dynamics.

The rise of big data means that citizens and academics can access much more information than in the past. Back in the 1990s, a large county such as Los Angeles might have twenty-five smog monitoring stations open and operating. These stations might record the daily high level of ambient air pollution. Many environmental justice activists have worried that too little data is collected in high-poverty and minority areas. Issues of data manipulation have arisen in China as its government has been accused of reclassifying days as "blue skies" that were actually polluted.[20] The rise of cheap, continuous-time air pollution monitors allows individuals to collect their own data about pollution, and this encourages accountability because such information can be used in class-action liability lawsuits.

Other examples of big data creation are more humorous. In July 2018, the *Wall Street Journal* reported that the professional basketball player

Jeremy Lin had invested in a company called HomeCourt whose core product uses an Apple iPhone to film a player shooting a basketball. From this video alone, the firm's algorithms crunch the resulting data to identify spatial and temporal trends concerning what types of shots a player makes at what place on the court.[21] These probabilities are useful for seeing data patterns and identifying areas that require improvement. Big data is now more easy to create (the computer transcribing a video into a matrix of data). In the past to create this Jeremy Lin data spreadsheet would require that a person watch a video of Lin and transcribe notes and type them into a spreadsheet. By lowering the cost of processing information, sports players and coaches can quickly learn from a huge amount of data how to improve a player's performance.

This sports example demonstrates how athletes use big data findings to hone their craft. Academics are following a similar path by partnering with cities and private companies that create these data. Some of these studies are related to climate adaptation. For example, on extremely hot days, do all households increase their electricity consumption or do only richer people?

The big data revolution helps the public sector to protect people. During heat waves, the local government can map out the spatial geography of where 911 emergency calls are originating and where crimes are being reported. Such information helps the police and ambulances to prepare for where to station their officers. Using geocoded data, researchers and government officials can identify those neighborhoods in the city generating the most hospitalization calls on extremely hot days. By directing additional resources to such areas, such research reduces the health impacts of such events. For example, if people who live in buildings built between 1935 and 1950 are more likely to suffer heat stroke on extremely hot days, local authorities can locate public cooling centers near such areas to help residents cope with extreme conditions.

Air conditioning protects people from extreme heat. Real-time indicators of electricity consumption provide direct clues on who is and is not engaging in such averting behavior. If researchers can access an electric utility's residential electricity consumption data, then they can collect data for each residential customer. If the data are available each hour, they would observe the home's zip code, the day, the hour, and the total electricity consumption for that hour under that account number. By knowing the account number, they can compare the same household's hourly electricity consumption on hot and cooler days. Since the researchers know what day and hour the electricity meter was operating, they can merge in weather data for that day and hour. Because they know the household's zip code, they can use census data to identify rich and poor neighborhoods in the city. If there are zip codes where electricity consumption barely increases on hot days and if these zip codes are poor, then this suggests that these areas are not using air conditioning to adapt on these days. This finding would have public health implications.

Once climate-vulnerable populations are identified, interventions can be piloted to improve their quality of life. Similar to a medical trial with a formal treatment group and a control group, the set of vulnerable communities can be randomly assigned to receive a treatment while the other group would not receive the treatment. In a medical clinical trial to reverse baldness, the treatment would be a pill to cure baldness. In the case of climate change adaptation, the treatment might be to receive a free air conditioning fan. If the fan costs $100 and it generates $700 in total health and productivity benefits for the person, then this treatment clearly benefits the average person. The necessary data here would result from tracking the well-being of people in the control group and the treatment group over time. If such a randomized pilot program proves to be effective, then these findings can be broadcast through YouTube videos and other social media, and this program will be more likely to be scaled up in other locations for other poor people. Good ideas are public goods. Once we discover them, we can introduce them in many places and benefit many vulnerable people.

A Case Study of Natural Disaster Adaptation Based on Big Data

In September 2017, Hurricane Maria greatly damaged Puerto Rico. In the aftermath of this terrible hurricane, many people have left the island. Researchers have studied this migration by using cell phone records. They looked at cell phone records to determine who was living in Puerto Rico before Hurricane Maria. They then looked at subsequent phone calls made months after this hurricane, along with information on the location of the closest cell tower, to establish the new location of the cell phone holder. By using cell phone location data for the same person over time, researchers can study how displaced climate refugees choose a new place to live, such as Dallas or Houston. What role does the destination's unemployment rate and the percentage of people in this area who are of Puerto Rican descent and have already settled there play in determining the likelihood that a Hurricane Maria refugee moves there?[22] Research based on cell phone data will not give the researcher demographic information about individual cell phone users. Is this person a man? If so, is he young? Is he well educated? Without such microdata, the researcher has a limited ability to test hypotheses about variation across demographic groups in migrating when a disaster occurs. Such data also do not provide direct evidence on the person's quality of life. We know that the person is still alive and we know his location, but we do not know his income, whether he is healthy, whether he is employed, and whether his family is with him. If government can access such data, then it can text such individuals to help connect the new migrants to basic social services. In this sense, the big data revolution helps individuals to recover more quickly from climate shocks.

The Payoff of Combining Cross-Platform Big Data

Researchers learn distinct lessons from the content posted to different Internet platforms. Consider what a researcher learns from Twitter social media content versus Google searches if both are conducted by the same person. One innovative research team used billions of tweets to establish that when people are repeatedly exposed to extremely hot weather, they tweet less about subsequent heat waves.[23] The authors interpret this as evidence of a boiling frog effect. They argue that we are the frogs and that, by not tweeting about extreme heat after prolonged exposure, we have revealed that we are unaware of the increasingly terrible conditions we face. In their interpretation, we become inured to the heat. An alternative explanation is that people do not tweet everything they feel. People may have a taste for novelty, and if they feel they have repeatedly made the same point in previous tweets, then there is no point in tweeting on the topic again. In this case, Google searches are more likely to be a better mirror of our soul than our tweets. When we express ourselves on social media, we are aware that this is not our own private diary. We recognize that tweets are for public consumption, whereas our Google searches are a private activity.

This simple point leads me to conclude that both sources of information provide valuable clues about a person's priorities at a point in time. In fall 2018, I submitted a proposal to the National Science Foundation to compete in the "Big Idea" for the year 2026.[24] I proposed that a representative sample of Americans be enrolled in a program such that their daily Internet interactions be recorded in a standardized data set. To be specific, if I was a member of the sample, then each of my Google searches, Amazon searches, cell phone text messages and my location (based on my phone), tweets on Twitter, Twitter searches, and Facebook posts would be recorded in a standardized database that would include the geography of where I was as I interacted with the Internet and the time when these interactions occurred. This integrated platform data tracking the same individuals (who would not be identified by name) over time would provide invaluable insights regarding how people cope with stressful times. By tracking the same individuals for a year, researchers would see how people use social networks and search engines such as Google to respond to emerging threats such as heat waves, fires, and natural disasters. Such geocoded data could be linked to outdoor temperatures and air pollution levels to study how such external environmental factors are correlated with Internet platform interactions. The final link here would be to connect this back to real-time indicators of a person's health such as their heart rate, sickness, and discomfort. This would be the final outcome indicator of a person's ability to mitigate daily evolving threats he or she faces.

My NSF proposal was rejected! To appreciate the value of such a data project, recall that in December 2019 in Wuhan, China, the covid-19 virus began to spread and some doctors there were worried that a true pandemic was playing out. Imagine a world in which China allowed Google to operate there and the people of Wuhan used Google to search for health clues about the conditions they were experiencing. Google Trends has used spatial correlations in its searches to identify flu outbreaks.[25] Google data scientists would have spotted an early rise in contagion and could have notified worldwide authorities. Local leaders in Wuhan would not have been able to cover up these facts, public health measures would have been taken earlier, and the contagion would have been slowed. The cross-platform data embodied in my proposal would have allowed for a greater understanding of which people were susceptible, and this could be achieved without physically examining anyone. Such low-cost remote testing would have accelerated pinpoint efforts to isolate those at risk.

Big Data Property Rights

This ownership structure of our big data may slow the rate of progress concerning our understanding of how different people respond to emerging climate conditions. Amazon has strong incentives to hire data scientists to work within the firm to help it to make a profit. It has much weaker incentives to work with academics to help them further their climate resilience research. Facebook faces major regulatory challenges for its sharing of data with firms such as Cambridge Analytics. Corporate lawyers have always worried about the damage that data sharing poses for firms. The recent headlines only accentuate this issue, making firms more hesitant to share data with academics.

Consider Google's proprietary data. As each individual engages in a Google search, the individual learns from the wisdom of the Internet and Google learns more about this individual. Google is a profit-maximizing firm. The pursuit of profit aligns with increased climate resilience if Google targets the searcher with advertisements that both generate profit for Google and help the individual to achieve a goal. For example, if the individual seeks out mold repair specialists, Google has profit incentives to prioritize specialists who might sign a mutually beneficial deal with the searcher. In this sense, Google's search engine facilitates matching to solve new problems that Internet searchers face. While Google's profits are increased in this fashion, it gains less from sharing its data with academic researchers.

A partnership between Google and academics would accelerate adaptation research progress. Google can geocode and time stamp each web search and knows your IP address, so it can follow the same people over time. As each person searches for different pieces of information in real time, Google learns about a huge number of people's perceived threats and opportunities. Google has used this information to study the business cycle (people searching for jobs or unemployment) and to predict epidemics.[26] If Google would partner with more academics on research projects focused on understanding search behavior dynamics, then adaptation research would accelerate. It seems that Google has only weak incentives to engage in such partnerships.

The Amazon platform would provide invaluable information concerning how people budget for emerging challenges. A weakness across these platforms is that they do not reveal a family's budget sheet. Credit reporting agencies such as Equifax know each household's credit score, and this information, if it could also be merged at the individual level, would allow researchers to track how poor, middle-class, and rich households cope with emerging challenges.

Big data provides a real-time thermometer focused on how different people cope with the same heat, pollution, and disaster risk challenges that affect a given area on a given day. Such information can be used to test how able richer and more educated people are to reduce their losses from extreme climate conditions. To create this special data set would require these profit companies to share some of their proprietary data. How society creates incentives for such open-source learning using private sector data remains an important near-term issue.

8

Reimagining the Real Estate Sector

Roughly 63 percent of Americans own their own home. Home buyers make a major place-based bet. I own a home in Westwood, Los Angeles. I have bet that my neighborhood and the greater Southern California area both have a bright future. Even though standard financial advice recommends diversifying one's investments to avoid putting all of your eggs in one basket, many people ignore this advice. In the midst of rising climate risk, this is an increasingly risky bet.

The Rising Benefits of a Renter Economy

Renting offers a number of adaptation benefits relative to owning. Renters are more geographically mobile because they face lower migration costs. Homeowners with a mortgage must arrange to sell their home to pay off the mortgage. Home ownership raises people's unemployment risk because they are tied to a specific geographic area.[1] If a local labor market such as Chicago suffers a major recession, both local wages and rents decline at the same time. A renter will have an easier time migrating to another local labor market that offers greater opportunities. By making fewer place-based bets, renters hold a more diversified financial portfolio.

By the definition of diversification, place-based shocks (such as Houston's Hurricane Harvey) will have minimal impact on the portfolio of someone who has invested in a diversified mutual fund. Having most of your life savings tied up in housing wealth contradicts standard diversification advice.

The home buyer is betting that the local labor market will stay strong and experience job growth. The buyer is also betting that the local government will continue to deliver good public schools and safe streets. The purchaser is also betting that the area will be resilient in the face of changing climate conditions.

Whereas the United States has a tradition of home ownership, other nations feature much higher rates of renting.[2] Home ownership has been a key piece of the American Dream. Moving forward, such a status symbol can be replaced with other ways to achieve status. It remains an open question why pride of ownership is so important in US culture.

If more individuals are renters, this offers private benefits to them, and it offers social benefits in terms of society having a more stable financial system. During the 2008 financial crisis, middle-class people were more likely to default on loans. A homeowner's likelihood of defaulting on a mortgage loan increases if the homeowner is unemployed and owes more on the mortgage than the current value of the home.[3] Middle-class people are more likely to default because they both face a greater risk of becoming unemployed and often have less housing equity, since they make smaller down payments when purchasing a home. Such defaults can set off a domino effect in one's neighborhood. If your neighbors have recently defaulted on their loans and if the bank turns around and quickly tries to sell this unwanted collateral at a fire-sale price, then this will depress the value of your nearby home and make you more likely to default.[4]

Increased climate risk could increase mortgage default risk. Real estate owners are more likely to default if the current market value of the home is less than the current mortgage on the home. Homeowners will be less likely to default if banks that lend them money to purchase a home require larger down payments. In this case, owners will have more equity in the house and will have less of an incentive to default on their mortgage balance. For example, suppose a homeowner buys a home for $400,000 and purchases it with a $40,000 down payment and a $360,000 loan. The small 10 percent down payment means that this buyer has little equity in the home. If due to a negative shock, the home is now worth $340,000 and remaining loan due is $350,000, then this owner now has negative equity in the home equal to –$10,000 and has an incentive to default. Now consider the same example but suppose the buyer had to finance the house with a 50 percent down payment of $200,000. In this case, even when the home's price has declined to $350,000, the homeowner still has $150,000 in housing equity ($350,000 minus $200,000) and will not default on this loan. This example highlights that an alternative to encouraging renting is to require that new home buyers make a larger down payment for a home. By lowering the loan-to-value ratio, the lenders would face less default risk.

Three Risks Associated with the Renter Economy

In booming areas, renters can face large increases in their rents. As such an area gentrifies, some people may no longer afford to live in their old neighborhood. Owning insulates people from this risk.[5] In cities such as Boston and San Francisco, local land use zoning policies contribute to rising rents. In these coastal cities, both the terrain and the urban politics are such that it is very difficult to build new housing. If real estate developers could build more housing in areas where demand increases, then this would limit the increases in local rents.[6]

Renting also poses a risk for the owner of the asset. During a time of rising income inequality, many progressive cities are strengthening rent control laws and antieviction laws. In cities such as Berkeley and Santa Monica, and perhaps soon in Los Angeles, landlords face limits on rent increases. These price controls lower the probability that renters are priced out of their units. But the limits implicitly represent a taking of the stream of revenue as it is transferred from the landlord to the tenant. As some cities make it harder for property owners to evict tenants, some renters will be emboldened to not pay their rents. In this case, the owner will collect less revenue and earn a lower rate of return on this investment.

Suppose you buy a million-dollar home with the intent to rent it out for $60,000 a year, or $5,000 a month. If after you buy the home, local voters enact a law that states that rents cannot exceed $4,000 a month, then it is likely that you will regret buying this asset. In this case, you are required to continue to rent the property out at an artificially low price. In the language of economics, the opportunity cost of this is that you could have taken that original million-dollar investment and put it in the bank and earned interest. Property owners do face political risk that local officials will raise property taxes or limit how much they can collect in rent. If investors anticipate these risks, then they will bid less aggressively for housing assets and the current owner will suffer an asset value loss. A second risk that property owners face is damage to the rental property. In a rental contract, renters are the people who use the asset every day, but they do not own the asset. This separation between ownership and control creates bad incentives. For example, renters may have little incentive to take care of the property because they plan to live there for a short time. The rise of big data mitigates this challenge. Property owners can now more cheaply monitor their tenants using drones, remote sensing, satellite photos, and real-time access to electricity and water consumption to check in on how their asset is being treated. This information helps the asset owner to know in real time whether a given contract's terms are being honored.

A third risk posed by shifting to a setting where more people are renters is that community civic-mindedness might decline. Some leading scholars have posited that home ownership contributes to local social capital.[7] Urbanists have documented that home ownership represents a long-run investment in a specific geographic community. When people expect to live in a location a long time, they are more likely to socialize with their neighbors and to take actions to make the local community stronger (neighborhood policing, calling local representatives about emerging issues, and so on). They will devote more effort to keeping the property well maintained. In our big data economy, the owner of a rental property will recognize that the tenants do not have strong financial incentives to maintain the property or to be civically engaged. The owner can adapt to this issue by investing in basic property maintenance and by holding events to bring various community members together. Just as the owner of a shopping mall seeks to maximize the positive synergies across shops at the mall, so the owner of a large number of nearby rental properties can encourage a local community to flourish through events such as picnics and block barbeques. Such events can proceed even if everyone is a renter if the owner of the property anticipates that there is a demand for creating a community. While this may appear to be a minor example, it actually dovetails with this book's main theme. In capitalism, when decision makers anticipate an emerging challenge, there are always strategies available to mitigate the challenge. This dynamic means that the social cost of the challenge can be reduced.

Homeowners as Place-Based Public Policy Lobbyists

Homeowners have strong incentives to lobby state and federal elected officials for local place-based policies that enhance their assets' value.[8] Such lobbying can be socially beneficial if officials enact public goods policies such as reducing air pollution and crime using regulations that are cost effective. Such lobbying can be socially costly if such lobbying seeks to obtain local subsidies at the expense of general taxpayers. The existing rules actually hinder climate change adaptation while boosting the asset values of properties in areas facing fire, drought, and flooding.[9]

For example, the federal government's expenditure for providing fire protection in the wildlife-urban interface encourages more development in fire zones.[10] Today, the federal government subsidizes fire protection.[11] In recent years the US Forest Service has spent over half of its budget on fighting wildfires. This share is expected to rise in the coming years. This agency is spending more than $1 billion annually to fight fires.

Purchasers of real estate facing risk have strong incentives to be aware of what government policies exist to protect them from this risk. A prominent example is the federal government's National Flood Insurance Program (NFIP). If one's residential community has enrolled in the NFIP, then a homeowner is eligible for $250,000 in damage insurance after spending roughly $700 a year. In a world facing climate change, this is an increasingly good deal for homeowners in eligible areas.[12] Given that the government offers this subsidized product, private insurers have fewer sales opportunities to market profitable flood insurance.

FEMA's subsidized flood insurance acts to implicitly subsidize development in flood zones. After Hurricane Harvey hit Houston in 2017, FEMA spent $11 billion in insurance claims in meeting its obligations under the NFIP.[13] This level of subsidized insurance helps homeowners to recover from horrible disasters. This ex post access to funds is very valuable. Yet economists worry that access to subsidized insurance creates bad incentives. If homeowners anticipate that they will be bailed out by the federal government if a low-probability terrible event occurs, then this diminishes their incentives to take a variety of precautions. If these individuals knew that they had no access to this safety net, they would either be less likely to live in risky areas (and the big data revolution helps to identify these) or build their homes using more resilient materials.

If residents were renters rather than owners, then Congress would be less likely to be generous in subsidizing insurance. Homeowners who live in a specific jurisdiction are at once residents, voters, and asset holders. They have strong incentives to pursue spatial subsidies that benefit them. Absentee landlords tend not to vote in the jurisdictions where they own property. Although they can make campaign contributions to politicians, they are unlikely to have the same clout as large numbers of homeowner voters.

Residents who rent properties in increasingly risky areas would not lobby as hard for subsidies to protect their areas. Renters would recognize that since they do not own the house they rent, they do not gain an asset value boost by lobbying for government largess. By decoupling the pursuit of spatial subsidies in an increasingly risky place (due to climate change) from voting and lobbying activity, the rise of the renter economy would reduce inefficient lobbying activity.

Under the current rules, incumbent homeowners located in risky places have strong incentives to make the case that they are naive victims who need government insurance and protection. This has several ugly consequences. First, it encourages people to live in the increasingly dangerous areas.[14] Second, given that governments face balanced budget constraints, this means that other people's taxes must be increased to pay for the bills incurred for firefighting in fire zones and offsetting flood damage in flood zones. These higher taxes imposed on the people who choose to live in relatively safer places both lowers their after-tax income and distorts their incentives, encouraging them to save less and work less because they keep less of their after-tax income.

A third consequence of the active role that the federal government plays in subsidizing insurance in risky places is that this displaces the private sector's insurance efforts. The private sector insurance industry has incentives to pursue profitable insurance contracts. If the federal government stepped away and stopped subsidizing insurance, some risk-averse homeowners would choose to be renters in such areas and other risk-averse people would seek private insurance. The private insurance industry would invest more in big data and spatial risk delineation and hire expensive talent to evaluate these risks if there is a growing set of customers seeking to purchase such private insurance policies. Since the federal government insures these areas, the private sector has less of an incentive to invest in this expertise.

Developing Resilience Skills through Human Capital Investment

Underlying the economist Julian Simon's optimism about the ongoing improvement in our standard of living is the ever increasing investment in human capital. Such skill builds on skill, allowing us to make new discoveries. This general insight has key implications when thinking about the consequences of encouraging a transition away from home ownership and toward more people renting housing.

Individual homeowners are amateurs at managing their properties. While some trained engineers and home improvement hobbyists have the skills to upgrade their properties, the typical owner relies on local repair people. If fewer people own the home where they live, then somebody else will own this asset. Real estate professionals with expertise in managing properties have an edge in managing such groups of nearby homes. Consolidation of ownership by a real estate holding company would mean that this company would have both expertise and access to the financial capital to invest in resilience investments to protect the real estate assets. Although the specifics would vary case by case, putting coastal homes on stilts would be one example.

New rules for promoting the rise of the renter economy shifts the management decision of a home from an amateur homeowner (who may underestimate risks and not know how to respond to them cost effectively) to a professional, publicly traded company with boards of directors watching over it. In this sense, the renter economy offers more accountability and more professional management of this key sector of the economy.

Real estate investment trusts, or REITs, are publicly traded companies whose assets are collections of real assets such as hotels or commercial office buildings. If more residential properties become rental properties and if these homes are bundled into REITs, then the major shareholders of these REITs have strong incentives to monitor the real estate company's management to make sure that these executives are hiring firms such as Jupiter and Coastal Risk Consulting to stay ahead of new risks that their properties face. Major shareholders in REITs act as the adults in the room because they could lose money if the REITs underestimate new risks.

The large real estate developers who develop new construction have the financing, human capital, and economies of scale to supply high-quality housing. If private companies own the properties in the wildlife-urban interface, they would have the incentives and the resources to provide their own fire protection for such areas without public funds. If a private developer did not provide fire protection, such a property's riskiness would be common knowledge (due to the rise of climate big data firms), and this extra risk would mean that the rental prices would be lower to compensate people for the extra risk they are taking by living there. The profit-maximizing suburban-rural developer would install fire protection if the extra rents it could collect by having a reputation for having a safe, low-fire-risk property exceed the extra costs it incurs by engaging in this self-protection. In such a planned community, the developer has strong incentives to internalize all social costs of emerging risks.

Research on the economics of shopping malls and on suburban developments have both documented this fact. The owners of these complexes have strong incentives to encourage positive synergies within such a complex (such as having green space or a nice atrium at a mall) and incentives to combat negative externalities (such as fire risk).[15]

In a renter economy, some real estate developers might worry that they hold an undiversified portfolio if all their money is tied up in managing location-specific real estate such as a condo building in Miami. Such real estate developers would have the option to sell such buildings to a REIT. Less risky assets will sell for a higher price. If a REIT earns a reputation for having climate-resilient real estate in potentially risky locations, then its properties will be occupied and rents will be high.

Real estate management companies would have strong incentives to work with risk prediction teams such as First Street Foundation to measure the risk they are exposed to. Such real estate holding companies could diversify this risk by purchasing catastrophe bonds. Catastrophe bonds are a financial instrument designed such that if a specified disaster occurs, then the holder of such a bond must pay the victim a lump-sum amount of money. If no disaster occurs, then the bondholder receives a flow of payments. This financial asset acts as a type of insurance policy guaranteeing the bond seller that if something horrible happens, the victim will receive funds. Real estate management companies could also transform themselves into REITs and sell shares in their companies to outside investors. This approach would allow for the place-based risk to be diversified across many investors' portfolios.

Footloose Renters and Charles Tiebout

Charles Tiebout is famous in urban economics for his work on how local governments compete against one another.[16] Each geographic location has natural features. For example, Berkeley, California, has great weather and beautiful hills and bay views. Taking these features as given, local elected officials choose what services and taxes to provide. Some will choose to offer low-quality services and ask residents to pay little in taxes. Other areas will have great public schools but have a high tax price.

Tiebout argued that a diverse population will differ with respect to their rankings of which communities they want to live in. Some will seek out areas featuring low taxes and low local public services because this package allows such households to have more after-tax income to spend on goods and services they want to buy. Those people with a taste for high-quality public services will recognize that high taxes are the price of providing such services. Tiebout argued that communities choose where on this tax and services spectrum they wish to locate. Communities in Florida feature low taxes and low-quality services, whereas California is known for having higher taxes and higher levels of public services such as excellent public universities. People will vote with their feet and choose the best community for them.

Some communities will use their tax dollars to invest in fortification along a number of dimensions such as flood control and planting trees to offset the heat island effect. A society that embraces the Tiebout vision accommodates many different beliefs and priorities.

A valid concern is whether the poor can afford the rent in the safer communities. If rents are much higher in safer communities, housing vouchers could be issued to increase this group's access to safe areas. Alternatively, as I discuss in the next chapter, changes in the zoning code could be permitted to increase the supply of housing in safe areas. In a renter economy, the incumbent homeowners who already live in an area would be less likely to oppose new housing construction.[17] If renters living in a low-service, low-tax area learn that climate change is worse than they thought, they still have the ability to move to a more expensive, safer area.

Would housing suppliers of these rental communities advertise that their planned community is resilient but lie about this in order to save on costs? Economic theory makes two predictions. First, firms value their reputation. They will recognize that, in this Internet-connected world of Yelp reviews, the production of shoddy products will cost them future sales if they are caught overstating the resilience quality of their residential communities. Second, the federal government could play a role similar to the Department of Energy's Energy Star program certifying the resilience quality of such communities.

Encouraging Resilience Skill Investment

Middle-class homeowners may not have the savings to upgrade their homes to make them more resilient and thus engage in deferred maintenance. Such homes will be less resilient in the face of extreme climate conditions. Consider a neighborhood featuring two hundred homes. Rather than having two hundred separate homeowners who are amateurs running these structures, imagine if one real estate company owned all two hundred homes and rented them out to those families. This company will have the right incentives to pay the fixed cost to hire the best engineers to work on emerging problems. This demand for such solutions will create incentives for engineers to accumulate the human capital to excel in solving these problems. In prior research, I documented that Walmart is more energy efficient per square foot than the smaller retailers it competes with.[18] Walmart's scale of asset ownership gives it an incentive to hire the best talent to provide blueprints to be efficient, and these ideas can then be replicated across its many superstores. The same logic holds in protecting real estate from climate risk. The consolidation of ownership creates scale economies so that the right talent—and over time this talent is built up due to anticipated demand—is employed to solve increasingly tough problems. Each young person chooses what skills to acquire in school in order to have a productive career. As climate resilience challenges increasingly arise, more young people will specialize in developing such skills. Professional real estate companies will hire them to work on projects, and the net effect of this human capital investment will be higher-quality real estate structures that can withstand climate shocks.

Given that large-scale real estate owners can hire the best consultants, they can use their land more productively than an amateur homeowner. Such large-scale real estate owners are likely to have more cumulative expertise and can hire excellent consultants when they are confronted with a hard new problem such as mitigating flood risk. This productivity differential means that the at-risk assets are now controlled by a larger entity. As such, each square foot of real estate will now be safer. The important empirical question here is whether this consolidation of real estate that is managed by experts is sufficient to protect real estate fully from the climate shocks that may play out over the next decades. This claim can be tested based on natural experiments. As future hurricanes affect coastal areas, do they cause less economic damage in areas featuring consolidated ownership of real estate and cause more damage in areas featuring middle-class homeowners?

Encouraging Resilient Development

In a renter economy, the housing stock could be both of high quality and climate resilient if renters demand these attributes. Suppliers can charge a price premium for delivering differentiated products that consumers want. In the car market this is well understood. Safety concerns create a strong incentive for General Motors to build safe cars for people to purchase. If growing concerns about climate change lead renters to seek out safer housing, then this will become an attribute that developers will try to bundle into their housing and then charge a price premium for. An open question is how a real estate developer will develop a reputation for building safe, resilient housing. In the case of cars, Volvo has developed a reputation for building safe cars. There are a huge number of such cars on the roads driving each day, and a good statistician can collect data on whether people are more likely to survive crashes when they drive a Volvo. This is a direct test of whether these vehicles are safe.

In the case of home building, will there be an analogous natural experiment? As natural disasters such as tornadoes and hurricanes hit, if homes built by the national builder KB Home withstand these shocks, this is direct evidence that their product is more resilient as it can literally take a direct hit. The *Wall Street Journal* has reported that some real estate developers are exposing their new buildings to extreme hurricane conditions to see how they perform and then advertising the results from these tests.[19]

An alternative approach to building resilient real estate is to build less durable structures that are meant to be lived in for only twenty or fewer years.[20] These structures would be cheaper to build. They could be built out of modular materials that could be disassembled. If flooding in an area turns out to be worse than was previously expected, perhaps because of sea level rise, then the materials can be dragged to higher ground and reused. The owner of the property would have less capital at risk and holds an option to rebuild in the future as climate science makes more progress concerning the spatially refined risks that an area faces. If the area faces great risk, then less capital should be invested there.

At any point in time, most housing is several decades old. Older housing begins to age and need an upgrade. If apartments and homes are owned by large investors, will they retrofit them to upgrade the quality of such units? Energy economists have noted that real estate owners sometimes do not invest in energy efficiency upgrades. Real estate owners will be more likely to invest in a structure's climate resilience if they can charge higher rents. In the case of real estate energy efficiency, there is an issue of certifying that an apartment is energy efficient. In Holland, a system of certification has been introduced, and research has documented that the energy-efficient structures sell for a price premium.[21] If the building owner could charge higher rent for a more energy-efficient apartment (or if the building owner pays the tenant's utility bills), then this would create an incentive to invest in more energy-efficient durables such as windows and dishwashers.

The same logic applies for achieving risk resilience. If there is a trusted certifier of green resilient real estate, then this quality certification would create an incentive to supply it. For a particular piece of real estate in a given location, a question arises concerning how we judge whether the building is resilient. If over the past thirty years it has suffered no flood damage, is that because the building is well constructed or because the building was lucky? For low-probability events such as extreme flooding, it is more difficult to disentangle these two explanations. A building owner has strong incentives to overstate the structure's ability to withstand extreme conditions. Civil engineers could create consulting companies and earn a reputation (similar to restaurants with Yelp ratings) to compete for the certification business.

An open question about the rise of new climate resilience real estate focuses on the time horizon. If everyone agrees that future climate change is a risk for new Miami condos but most renters believe the risk will really take place starting fifteen years from now, will a developer of a new building incorporate such costly resilience features? This developer may choose to build fewer housing units on the same amount of land to provide more green space if this green space simultaneously is a park amenity, provides flood control, and protects against the urban heat island effect.[22]

The Political Challenge of Phasing Out Tax Incentives to Own Housing

Throughout this chapter I have argued that there will be increasingly large benefits to our economy by reducing the count of people who are homeowners. This transition could be accelerated if the federal government changes its tax code and mortgage rules. The government has enacted rules that subsidize home ownership. One rule is that homeowners do not pay taxes on their imputed rents.[23] In other nations such as Switzerland, property owners pay such a tax. Consider homeowners who own a million-dollar home. If these homeowners did not live in this house, they could rent it out to another family and perhaps collect $40,000 per year. They would have to declare this income and pay taxes on it, but if they live in the house they do not declare this imputed rent as income. This tax savings is an implicit subsidy for homeowners.

Homeowners who finance their purchase using a large mortgage receive another tax subsidy. A household that has a million-dollar mortgage outstanding and has borrowed at 5 percent will pay $50,000 in interest for the year. If this household can deduct this interest from its federal taxes and if this household faces a 33 percent tax rate, then this household gains $17,000 annually in lower taxes.

Given that many adults live in owner-occupied homes, this group has strong incentives to preserve their current homeowner subsidies. Incumbent homeowners in high-rent areas (such as California) who have large mortgages benefit from the existing rules. Those who live inland and off the coasts benefit much less from these rules. Young people who pay taxes and rent benefit less from these tax rules. An open question is how a new political coalition could form to seek to reform the housing subsidies. In the aftermath of the covid-19 pandemic, the United States will be facing an even larger fiscal budget deficit. Such deficits eventually have to be paid back by some combination of raising taxes and cutting expenditures. A phaseout of existing subsidies for home ownership could be one part of such deficit financing.

9

Reimagining Laws and Regulations to Facilitate Adaptation

There is great concern about inequality such that the rich have access to the very best of everything market capitalism has to offer while the poor have limited access to such goods. Poor people will have more opportunities to adapt to climate risk if more people can live in safer areas. Climate scientists must identify such areas, and policy makers must enact a set of rules that allow more people to live there. In this chapter I propose a set of new local- and state-level rules that together will significantly increase resilience. New rules covering urban land use, urban transportation, and resource pricing together will create many new adaptation strategies for both poorer and richer people.

Urban Zoning

Land use zoning policies determine what types of buildings can be built in different areas. Current real estate zoning laws that set aside much urban land for single-family homes have the unintended consequence of pricing poorer people out of safer areas.

In the case of housing, zoning, and land use, regulations limit supply. This raises home prices in desirable areas, such as in downtown San Francisco, and in the near future it will inhibit climate change adaptation. We could achieve greater safety for more people if we use big data to pinpoint higher ground and have flexible rules for what housing can be built there. If the status quo land use policies remain, then there will be a limited supply of safe slots, and under a market system their prices will be high and the poor will not be able to afford them.

In Los Angeles, two-thirds of the area is zoned for residential housing, and 75 percent of that area is reserved for single-family homes and duplexes. Absent such local zoning codes, climate scientists would identify parts of such a city that face less high temperature risk and less flooding risk, and real estate developers would build taller buildings in such areas to economize on scarce land. Zoning artificially blocks these mutually beneficial gains to trade among real estate developers and people who seek to live in these cities.

Zoning regulation is an example of a key local government policy that determines how we adapt to place-based risk. Such regulation introduces a misallocation of land resources. In the United States today, the most productive cities, including Boston, New York, Los Angeles, San Francisco, Seattle, and Portland, have all enacted regulations to limit the quantity of new housing construction. Research in urban economics argues that such local zoning in productive cities reduces the overall macroeconomy's level of income growth.[1]

Given that these cities tend to be in cool, temperate places along the coasts, the imposition of housing supply constraints deflects the population to less temperate, hotter places (think of Houston, Phoenix, and Las Vegas), where they are likely to be exposed to more extreme heat days. In my past work, I document that this displacement effect unintentionally increases the carbon footprint.[2] This consequence is due to people driving more and using more electricity in these hot, sprawling cities than if they lived in cooler, denser cities.

Homeowners have a vested incentive to block new development. This raises the price of their asset, making them richer.[3] Incumbent homeowners often argue that new housing construction tends to congest their neighborhoods and that high-rise buildings would change the character of their area and block their views. If homeowners held the option to sell their land to a developer who could consolidate adjacent plots, this developer could build high-rise buildings on the combined parcel. In this case, the developer would be willing to pay more for the land, and the homeowner (who owns the land) would own a more valuable asset. Under current rules that combine fixed zoning for land (as single-family homes) and local control of zoning, incumbent homeowners have strong incentives to discourage new construction on vacant parcels in their area.

The city of Minneapolis is acting as an early guinea pig here. In October 2019, its city council approved the 2040 comprehensive plan. Under this plan, Minneapolis abolishes single-family zoning and allows duplexes and triplexes to be built anywhere in the city.[4] As this city enacts this policy reform, researchers will study how quality of life and housing affordability evolve over time in Minneapolis. This experiment will offer valuable lessons to other cities about the intended and unintended consequences of upzoning.

During an era of low urban crime, physical proximity among people becomes a valuable attribute. While the ongoing covid-19 pandemic poses risks to dense cities such as New York City, it is highly likely that more people will want to live in multifamily apartment buildings that are conducive to an urban lifestyle.

During times of high crime, private lots have value because they offer privacy and physical separation from neighbors. In cities such as Singapore, Hong Kong, Shanghai, and New York, rich people are willing to live in apartment towers and then use common green space such as New York's Central Park. As crime has declined in major cities such as New York City, people are more willing to interact with strangers and to go out at night. In recent years, more people (especially the young) have been attracted to living at high density. The connectivity of the Internet raises residents' enjoyment from leisure in such cities as they can both quickly find ideal leisure opportunities and coordinate with friends to enjoy such opportunities together.

If more productive cities featured more multifamily housing units, people could rent in these areas and learn whether they enjoy this lifestyle. Over a lifetime, people can live at high density when they are young adults and elderly and live at lower density when they have school-aged children. Current land use patterns discourage such dynamics and thus allocate huge amounts of productive land to single-family homes. If we live at higher density, such land could be reallocated to alternative uses. Some of those uses, such as parks and wetlands, would facilitate flood control and offset the urban heat island effect.

Urban Planning and Zoning for Multifamily Housing

As the cases of Boston, Manhattan, Shanghai, and Hong Kong highlight, multifamily apartment buildings can cater to the wealthy. In cities such as Los Angeles, many homes feature large lots. If developers did not face zoning restrictions, some would purchase adjacent homes, knock them down, and then erect perhaps a five-story building in its place. This would create more housing and would raise density. As population density increases, this would spur demand for nearby retail stores that would improve the area's quality of life as a new urbanist walking space would be formed.

If more and more young people are willing to live in multifamily buildings, then this facilitates climate change adaptation because we can build future cities using less land. Consider the following combinatorics problem. If there are 340 million Americans and if they live at Hong Kong's density of 17,000 per square mile, then the urban population would need 20,000 square miles of land. The continental United States occupies 2.9 million square miles of contiguous land. This means that there is a huge set of possibilities for siting 20,000 square miles of land within this landmass. If just half of Montana was built up to Hong Kong's population density, 1.5 billion people could live there.

Some of these cities would need to be built from scratch. The construction of these new cities would create construction jobs for low-skill workers and would guarantee that cutting-edge technology is incorporated into the city's core infrastructure. Such cities could be used as guinea pigs to show residents and leaders of older cities what life could be like in these new cities. People in other cities could visit these new cities to see if they enjoy the lifestyle they offer.

Even within sprawling cities such as Los Angeles there are examples of new dense urban communities. The success of the Los Angeles Playa Vista community highlights that there is a demand for such a lifestyle. The wetlands area formerly owned by Howard Hughes for the construction of the famous *Spruce Goose* has been redeveloped as a series of new urban multifamily housing complexes.[5] Playa Vista is an informative case highlighting how land can be repositioned when land use regulations permit it. As people marry later and have fewer children, the demand for single-family homes on private lots declines.

Under current land use rules, incumbent homeowners often launch lawsuits to discourage new development. The red tape imposed on developers reduces their supply of new housing, and this means that middle-class and young people face higher prices for housing. In an economy featuring more renters, these resident voters would support zoning changes because their rents would fall as more housing is built. Assuming that climate risk assessment companies such as Jupiter can accurately identify higher ground (less flood risk), taller buildings would be built in these areas and overall resilience for the population would take place as more housing is built in the relatively safer locations within a city.

If more people live in multifamily housing units, this would facilitate the use of public transit as more people live in walking distance to fast public technology (think of New York City's subways). Urban planners support upzoning near public transit network nodes. For example, when I lived in Los Angeles, I rode the new Expo Line to USC. Culver City is one of the station stops along the line, and much of Culver City features single-family homes in walking distance to the new train stations. Urban planners argue that if the nearby station-stop areas could be rezoned for multifamily housing, thousands of people could adopt a low-carbon lifestyle by walking to this train. This dynamic would lead to a re-creation of many Los Angeles neighborhoods featuring more affordable housing and a denser city consisting of millions of people living in multifamily apartment buildings as renters with walking access to transit.

There are benefits of living at higher density. People will walk more and some may lose weight. Urban planners claim that suburban sprawl is a major cause of obesity, but economists have questioned this claim.[6] The big data revolution has improved crime protection by allowing the police to both respond and allocate their time efficiently. This encourages more people to go out at night, and this demand encourages urban entrepreneurs to offer an exciting supply of leisure and nightlife activities. The net effect is more vibrant neighborhoods.

A denser city would offer resilience benefits. Ecologists have documented that a benefit of preserving land as wetlands (rather than as housing) is to increase flood protection for the entire area. When new housing is built on wetlands, the roads and other infrastructure reduce flood control for everyone else who lives in the geographic area. In an economy featuring less stringent housing zoning codes, real estate developers can use climate forecasts to identify relatively safer areas and build high-density housing there. In an economy featuring 3D printing and prefabricated housing construction, we can actually build new cities much faster now than in the past. In Berkeley in 2018, a four-story building was built in four days using prefabricated materials.[7]

Reducing Development in Fire Zones

At a time when more fires are burning in the American West, fewer people will move to fire-prone zones if more affordable housing is available closer to such cities as San Francisco, Seattle, and Portland. The possibility that center-city land use regulation contributes to suburban growth becomes even more relevant as cities enjoy a reduction in crime and improvements in their vibrancy. As crime declines and with improvements in urban apps that connect people to exciting events, to restaurant activities, and to meeting with friends at these activities, more people want to live in such center cities. This demand drives up rents when little new housing is supplied. Changes to zoning codes would accelerate the construction of new center-city housing units.

If cities such as Boulder, Portland, and San Francisco feature more tall multifamily apartment buildings, this will slow housing construction at the fringe of such cities. In the West, there has been concern that more people are living in what is termed the wildlife-urban interface. The combination of drought and heat increases fire risk. The economic damage such fires cause is proportional to the count of people who live in the general area as they are exposed to fire risk and high levels of air pollution.

California's fire zone has been mapped out, and real estate developers and home buyers have strong incentives to be aware of this dangerous geography.[8] An open question concerns how much of housing growth in semirural areas is due to the desire to live in a high-amenity area (such as the forest) versus a deflection effect such that center-city zoning codes have raised urban home prices and pushed middle-class residents farther from the cities. In 2017, the economist Enrico Moretti published a piece in the *New York Times* in which he discussed the unintended consequence that center-city NIMBYism in California nudges more middle-class people to live closer to fire zones:

> The second negative consequence of the region's restrictive housing policies in the urban core is environmental degradation on the periphery. Good environmental stewardship suggests that we should build more in the urban core near transit and jobs and less on the fringes. Yet because of cities' strict housing regulations, we build more on farmland on the region's outskirts and less in the city center where demand is higher.
>
> Families who can't afford San Francisco, Berkeley or Silicon Valley have to move to exurbs. Some 3,800 Californians leave urban parts of the Bay Area for cheaper housing in Sonoma and Napa Counties every year. This worsens traffic and heightens the pressure for development on the edge of the region—in places such as Santa Rosa, home to some of the neighborhoods hardest hit by this month's fires.[9]

The US government provides subsidized firefighting services in many rural areas.[10] At first blush, this is the right thing to do because it protects people who happen to live in fire zones. Yet free firefighting services actually encourage more people to live there. This case raises key ethical issues related to risk exposure and risk perception. Consider an extreme case in which everyone who lives in the fire zone understands the real risks they face and recognizes that these risks are growing worse over time because of climate change–induced drought and heat. In this case, an economist would be confident to state that the free firefighting subsidies provided by the federal government are creating a moral hazard effect and that climate resilience would rise if these subsidies are phased out. Such a change in resource pricing (the pricing of firefighting services) would encourage more people to not live in the fire zone and, should they choose to live in such a risky area, to take more precautions to reduce their risk.

Now consider another case in which 95 percent of people who live in the fire zone do not perceive the risk while only 5 percent do. In this case, the vast majority of homeowners are ignorant of the true risks. Does society have an ethical responsibility to protect these individuals from their own mistakes? If the American people say "yes," then note that the

5 percent of the rational people are hiding out and benefiting from living with others who attract the generosity from the caring American taxpayers. This example highlights the issue of who bears the cost of climate change. If there are no federal subsidies for firefighting, then landowners in the affected fire zone bear much of the costs. If taxpayers provide the firefighting expenditure, then the costs of climate change to a particular place (the American West fire zone) are spread across the population.

Land use patterns and population density in a vicinity of fire zones in the American West would be different if homeowners had to pay for their own fire insurance.[11] Standard economic logic predicts that fewer people would live in these areas and that those who continued to live in such areas would invest more of their own money in fire precautions (such as their roof type and trimming of vegetation) if they knew that there would be no bailouts if a disaster occurs. Consider the following what if: if the government credibly declares that it will no longer provide fire protection expenditure, then households will retreat from the fire zone. This would mean that the same cause (a fire) would have a smaller ex post effect on causing damage both because fewer victims would be living there and because each remaining resident would have taken greater precautions.

This argument represents an application of the Lucas critique to climate resilience policy. Recall that the Lucas critique is named after the Nobel Laureate Robert Lucas. He argued in the context of the macroeconomy that as government changes the rules of the game, people and investors change their behavior. This means that one cannot naively extrapolate based on the recent past (when people made their choices given the past rules) to learn about their future behavior (when they face different rules of the game). Thus, observing a past correlation between fires and burned-down homes would not predict a future correlation because people have changed their behavior. Many fewer people would live in fire zones if we change two policies: urban land zoning rules and subsidized fire protection in fire zones.

Reforming Flood Insurance Laws

Access to insurance protects people against huge losses when rare disasters occur. Many economists believe that the private sector has an edge over the public sector in supplying insurance. In markets such as car insurance and life insurance, private companies supply such coverage. In the case of flood insurance, the National Flood Insurance Program offers subsidized insurance, and its existence has discouraged private insurers from researching and writing flood insurance contracts. FEMA's NFIP is now running a large deficit, and the program has only slowly been updating its flood maps to reflect the emerging challenge posed by climate change–induced sea level rise. If the federal government took the radical step of ending the NFIP and requiring that every homeowner purchase a minimal level of private sector flood insurance (with the option of buying more insurance), this would set in motion a series of adaptation-friendly events.

Requiring all property owners to hold flood insurance would mimic the recent approach pursued in health care insurance reforms. At the center of President Obama's health insurance reforms was the push to have more healthy people buy insurance to minimize adverse selection concerns. If only sick people purchase health insurance, then health insurers can go broke if they set the insurance premium prices too low. Healthier people are rewarded with lower premiums, but they are still buying insurance. A similar strategy could be pursued in the case of coastal flood insurance.[12]

If private insurers anticipate that there will be many flood insurance buyers, then they have strong incentives to invest in the human capital and expertise to pinpoint which pieces of real estate face the greatest climate change risk. Once these risks were diagnosed, they could write insurance contracts to reward homeowners for taking risk mitigation steps such as pruning bushes or putting the home on stilts. In the car insurance market, innovative insurance products have been introduced, such as paying insurance per mile of driving rather than paying a fixed annual fee independent of miles driven. Since all drivers must buy private insurance, the industry has strong incentives to research the demand for various products.

Such insurance innovation has not taken place in the case of flood insurance because the government has played such an active role through the NFIP. Subsidized public insurance has displaced the supply of private insurance, and it has discouraged the private sector from mobilizing the financial capital and human capital to truly study this emerging challenge.

A critic of my proposal would argue that homeowners in risky, flood-prone places will face higher insurance rates for purchasing flood insurance if the private sector replaced the NFIP's subsidized rates. This correct point raises two issues. Do the incumbent homeowners have the property rights to such a sweet deal? In the case of life insurance pricing, older people pay more for a life insurance policy because they face a higher probability of death in the next year as compared to a younger person. We will be better able to adapt to new climate change risks if the insurance industry is allowed to risk price real estate in a similar way that life insurance is priced for riskier policies. Put simply, why is real estate insurance treated differently than life insurance? A risk is a risk.

If we as a society truly worry about the distributional effects of raising insurance rates on incumbent homeowners who live in places where climate change has raised risks, we can grandfather in these property owners such that their rates would more gradually converge to the actuarial rates. While this will slow adaptation progress, it will increase political support for this policy reform.

Imagine the case where the government does not step in and subsidize home insurance in increasingly risky fire and flood zones. In this case, home buyers will bid less aggressively for housing in these places, and this lowers the profit for real estate developers who build new housing in such increasingly risky locations. When home buyers anticipate that they face the full cost of horrible events (such as floods and fires), they will only live in an increasingly risky location if the structure is built such that it can withstand these risks. In this sense, the tough-love, no-government-bailouts approach contributes to building up climate resilience because fewer homes are built in increasingly risky places and those homes that are built will be built with better materials to reduce their risk exposure.

Information Disclosure

George Akerlof won the Nobel Prize in Economics for his work on asymmetric information with the famous example of the market for used cars. The seller of a used car knows more about its quality than the buyer. In the case of used cars, the sophisticated buyer will delay the purchase and think, "Why is the seller selling? If the car is good, then would the seller be selling it? So, if the car is bad, I'm going to bid less aggressively for this car."

This lemons issue can arise in the face of new climate risks as the informed party (the owner of the at-risk asset) seeks to sell it to an unsuspecting potential buyer who is unaware of the risks. This scenario becomes more likely in real estate markets if the area is beautiful but increasing risky (such as coastal property). Potential buyers may be unaware of the emerging risks, love taking risks, or believe that federal government bailouts will protect them and insure them if a severe natural disaster occurs. An open question in the climate change economics literature asks, "How many real estate buyers know that they do not know the climate risk they are exposing themselves to by buying a given property?" For those sophisticated real estate buyers who know that they do not know the risk, they will conduct more research before purchasing the property.

North Carolina's coast offers a valuable case study on the lemons problem in real estate markets. In 2012, the state government passed a law that coastal sea level rise mapping must be based on historical data, not on future risks. The rationale for this North Carolina law was that the upgrades to the infrastructure that would be required if forward-looking maps were used would require large upfront expenditures, and that this would be a waste of people's money. Here, the North Carolina government is implicitly taking a gamble that North Carolina's coast can continue to prosper under business as usual.

If the federal government is worried that unsuspecting real estate buyers may be fooled into purchasing an increasingly risky property, it could require that all coastal home buyers take a thirty-minute short course to educate them about the risks they face. In California, state law requires that property sellers disclose information about the property's earthquake risk.[13] Similar strategies could be used to convey emerging fire risk and flood risk. Home buyers will be willing to pay less for those properties that face greater objective risk. Even naive investors have an incentive to do their homework before making a major financial investment. A prudent investment strategy in the face of such uncertainty would be to rent for a year in an area and learn more before locking in with a large down payment.[14]

If investors must make large down payments, then such individuals will have strong incentives to do their homework focused on a geographic area's flood risk and fire risk before seeking such a loan. If enough investors seek out trusted information, then this creates an opportunity for firms who can supply such information. On Wall Street, firms such as Moody's and S&P Global Rating play this role of providing third-party assessments of asset risks. Such information is reflected in the price of assets such that riskier assets have a lower price.

Currently, a prospective home buyer looks up the price and physical attributes of a property such as the year the home was built, its size, and its amenities. Zillow also provides information on walkability of the neighborhood, crime, and local schools' quality. As climate science progresses and more spatially refined predictions can be made, such local information can be expanded to include climate risks (temperature variation) and flood risk. This levels the information playing field and reduces concerns that there is asymmetric information such that potential buyers know less about the property than the sellers.

The federal government can play a key role as market maker if it requires that every home (even those that have not faced flood risk in recent years) must have some minimal flood insurance and fire insurance coverage. In this case, people shopping for homes will now have an incentive to research a specific home's flood risk. The aggregate demand for this information will create a market for firms that can provide high-quality reports about such risk. Those homes that are objectively riskier will sell for a discounted price only if the owner is unable or unwilling to take costly steps to protect the house against the emerging risks.

The nonprofit start-up First Street Foundation is creating a comprehensive database to mitigate this lemons issue in the case of property flood risk. I am a research adviser and a research collaborator with First Street because I am excited about the possibility of using the best climate science to accelerate climate change adaptation. First Street's comprehensive database shows whether homes in the United States have previously flooded or are at risk of future flooding. This information has previously been available to large investors but not to small investors and homeowners, who would face high costs to assemble such relevant data. The widespread dissemination of this information helps to reduce climate lemons concerns in real estate markets. As the foundation's webpage puts it, "For the vast majority of Americans, their home is their most valuable asset.

Understanding the true risk of flooding will enable homeowners to protect their investment."[15]

A legitimate concern is that the broad publication of flood risk scores on real estate Internet search platforms such as Zillow will lead to an overreaction such that home prices in flood zones decline too much. This could occur if home buyers in areas at risk for flooding are highly risk averse and if they have trouble understanding the risk scores that First Street publishes. In such a case, home buyers may overreact to the new news embedded in the risk scores. In this case, there would be an arbitrage profit available for those real estate investors who could buy several coastal homes at relatively low prices (because other bidders have been scared away). Such investors who own multiple properties would diversify the risk (by buying in multiple markets) and could wait to see whether the flood risk turns out to be less bad than the models predict. If a given coastal area that had been assigned a risky flood score does not suffer from a major flood for years, then some home buyers will be willing to pay a price premium for that home, and the arbitrager who has the capacity to hold homes in inventory (perhaps by renting them out in the meantime) would earn a good profit.

Transport Speed in Cities

Achieving higher speeds in cities would facilitate climate change adaptation because people will have a greater choice of neighborhoods. If Mary is willing to commute for thirty minutes one way and she can travel at 40 miles per hour, then she can look for a place to live in a 20-mile radius around her place of work. The area of a circle with a radius of 20 miles equals 3.14 × 400, or more than 1,200 square miles! Such a huge area provides ample opportunities for Mary to find the neighborhood that matches her desires. To appreciate this point, suppose that in a congested city, Mary can expect to commute at a speed of just 15 miles per hour. With a thirty-minute commute, she will be able to look for housing only in a 7.5-mile radius of her place of work, and this circle's area equals 3.14 × 7.5 × 7.5 = 176 square miles. This is a much smaller area to search over. By having a larger set of residential opportunities, each household will have a greater chance of finding a climate-resilient area.

In Singapore, due to road pricing, one can always expect to be able to achieve a speed of 40 miles per hour on the road.[16] While the rich are more likely to afford this, buses can also achieve these speeds, and with the economies of scale of a bus this lowers the per-person trip price for achieving this speed. The full cost of commuting includes not only the out-of-pocket expenditure on gasoline, parking, and road use fees but the value of the lost time. If a commute takes thirty minutes rather than fifteen minutes because of traffic congestion, then the commuter has lost fifteen minutes. Economists have adopted the rule of thumb of valuing such lost time by half of the person's hourly wage. For example, if I earn $80 an hour and I lose fifteen minutes stuck in traffic, then this costs me $10 in lost time ($.25 \times 80 \times .5$).

To conserve on such lost time due to congestion, cities such as Stockholm, London, and Singapore have adopted road pricing. Drivers in such cities move at higher speeds and save time but must pay more money out of pocket to travel at peak use times. One explanation for why so few cities have adopted road pricing focuses on behavioral economics: people are used to the roads being free. To an economist, this is a puzzling explanation because congested roads cost us valuable time. This time cost means that free roads are not free to use. A second explanation for the opposition to road pricing is that many poor people drive and they prefer to pay for their commute using their time rather than paying a road use fee.

In addition to phasing in road pricing, the rise of automated vehicle technology will facilitate speed increases both within and across cities. The fires in Paradise, California, highlight how road access during a disaster affects one's chances of escaping. Such individuals would have incentives to engage in ride sharing to increase the number of people in each vehicle.[17] By pricing road access, the government would have more revenue to upgrade roads and to build extra capacity. Such capacity would facilitate evacuations during disasters.

Imagine a future city featuring road pricing, automated vehicles, and widespread Uber-style ride sharing. In such a city, people would have greater mobility, and this would increase their productivity per hour. People would benefit from being able to access every part of the city. Because the transportation capital stock would be shared, there would be much less demand for private vehicles and parking spots because fewer vehicles would each be used more intensively. A huge amount of the land area in Los Angeles and other cities is devoted to parking. In this future city featuring automated vehicles and road pricing, people would rent vehicle services rather than owning a car. Because the cars keep moving around, there would be much less demand for parking. The concrete parking lots could be redeployed as housing and parks, and this would facilitate adaptation due to mitigating the urban heat island effect and offsetting some flood risk as more land could be set aside as wetlands. This example highlights how transportation innovation combined with new public policies (road pricing) allows for a scarce resource (land) to be redeployed to its highest and best use (some of it will be used for parks and flood control). By introducing economic incentives for how land is allocated, resilience is built up. The introduction of these new markets facilitates adaptation.

The Window of Opportunity for Enacting Pro-Adaptation Legislation

After a natural disaster, there is a brief window of time when new legislation can be enacted because the public is focused on the recent salient event. In a paper I published in 2006, I documented that a silver lining of environmental disasters is that Congress votes on more aggressive legislation intended to stop the next disaster.[18] An extension of this argument means that after major natural disasters, there is a brief policy window when interest groups can push for reform.[19]

While salient events may cause policy changes, it remains an open question whether such reforms are good or bad. Empirical work using cross-country data has documented that deaths from flooding decline in the aftermath of a major flood.[20] One explanation is that such salient events trigger adaptation responses that help to build up resilience. But there are other cases where salient events such as major fires lead to unintended consequences. Research documents that government officials who control fire zone resource allocation have overreacted to fire events and allocated too much money to the area that has experienced a fire. This outcome occurs because government officials are responsive to public pressure and the people are overreacting to the salient event. This well-intentioned allocation of funds can be misallocated if the money flows to a place that now faces less risk. Ecologists have noted that a silver lining of a recent fire is that such areas face less future short-run fire risk because there is less fuel to burn, but these are the areas that often receive more resilience funds.[21]

Ideally, society can avoid the latter case, but this hinges on learning from past mistakes. To learn from mistakes requires experimenting in many places across the United States to help us learn what works in building up resilience. Proresilience strategies will be more likely to be enacted if such policies can be shown to be effective in other settings. This point highlights the importance of policy experimentation and evaluation.

10

Innovation in Agricultural Production

Cowritten with Brian Casey and Nolan Jones

Agriculture is the economic sector that faces the greatest adaptation challenges due to climate change. As Rick Oswald, a corn farmer in Rockport, Missouri, put it, "I'm standing right here in the middle of climate change right now." Oswald's crop was ruined in spring 2019 by a bomb cyclone that caused unprecedented flooding in the Midwest and rendered twenty million acres unable to be planted. The storm destroyed four grain bins filled with corn on Oswald's property and flooded his family home for the first time in its eighty years of existence.[1] Farmers across the country have faced similar challenges: in June 2019, corn farmers in Ross County, Ohio, faced extreme challenges preparing to plant corn because of heavy rains.[2] In August 2018 in Sanpete County, Utah, farmers such as Scott Sunderland faced the challenge of extreme drought, which sharply reduced his farm's production of wheat and alfalfa.[3]

The US Department of Agriculture (USDA) paid out over $2.5 billion in crop insurance claims in 2019 but spent only $400 million on assisting farmers' adaptation efforts.[4] According to Oswald, he has never heard someone from his local USDA office bring up the subject of adaptation, much less receive funding for such efforts. Oswald's story captures the challenges American farmers will face in the future: an increased occurrence of extreme weather events that can lower yields or even prevent a crop from being harvested.

As the world's population and per capita income grow, there will be an increased demand for calories at the same time that farmers will face greater challenges supplying such output. If demand for food outpaces supply, then food prices will rise, and this will disproportionately erode the purchasing power of the poor, who spend a larger share of their income on food.

Research concludes that US agriculture has made little progress adapting to climate change.[5] Using county-year data for over fifty years to study the long-run relation between output per acre and outdoor climate conditions, researchers determined that corn's susceptibility to extreme heat has remained consistent. This suggests that little heat adaptation has occurred in past decades. Other research documents that there are certain critical temperature thresholds, such as 86 degrees Fahrenheit for wheat, above which the production of the crop declines sharply.[6] Given that climate change models predict that farming areas in California and the Midwest will face more of these extremely hot days, this portends future domestic food supply challenges.

Farmer skill, innovation, and experimentation could greatly help to address these emerging challenges. In July 2018, the *New York Times* profiled Jean-Baptiste Lécaillon, an entrepreneur making champagne in France. He has demonstrated a skill at growing grapes that can withstand the heat.

> Terroir and farming are of prime concern to Mr. Lécaillon, who took over as chef de cave in 1999 on the condition that he be put in charge of the vineyards as well. Responding to global warming and increasing the sense of place in the wines required some radical changes.
>
> He wanted the vines to have a much deeper root system that plunged into the bedrock of chalky limestone and clay; he believed that would help to protect against heat and drought while better expressing the character of the vineyard. To accomplish this, he eliminated the use of herbicides and fertilizers, developed techniques for training the roots downward and began trials for both organic and biodynamic viticulture.[7]

Lécaillon's actions anticipate several themes in this chapter. Those who are nimble in the face of changing conditions have a competitive edge in coping with climate change. The farmers who will lose the most profit from climate change will be those who are the least nimble to handle changing conditions. These individuals will have the option to sell their land to other farmers with an edge in handling the new conditions. Governments will need to adjust how they deliver aid to farmers in order to limit potential damage due to extreme weather events.

Adaptation Transition Paths

As our climate warms, outdoor farmers have access to at least three adjustment strategies: first, they can grow different crops more suited to their region's new climate; second, they can sell their land and purchase new land in an area suited to growing the crop they currently grow; and finally, they can devote their land to a form of activity other than farming.

In the American Southwest, climate change–induced drought will encourage farmers to consider alternative uses for their land. What becomes of this land will vary depending on the location of the farm: farms located near major productive urban centers such as Silicon Valley or Portland or Seattle will offer attractive parcels of land to real estate developers. The sales price of such farmland to developers will hinge on whether the zoning code can be changed to allow for smaller-lot housing. Internet companies such as Amazon continue to build warehouses at the fringe of major cities near transportation hubs. These warehouses are highly land intensive.

Today, California features extremely high home prices and is one of the nation's major farm producers. California accounts for over one-third of America's vegetable production and over two-thirds of its fruit production.[8] Much of the Central Valley—where the vast majority of agricultural production in the state occurs—is expected to turn into desert over the next hundred years, based on the existing climatic trends in the region over the past century.[9] Rural land market prices already reflect climate change forecasts such that there is a price discount for farmland where drought and extreme heat are predicted to worsen over time.[10]

Land can be converted to new uses such as suburban housing. Over time as urban transportation speeds increase (perhaps due to automated vehicles and road pricing), the suburban boundary will shift out farther from the city center and into farmland. Given that indoor structures can better withstand heat and extreme weather, this reassignment of farmland to residential and commercial air-conditioned real estate represents a beneficial adaptation transition.

Research in urban economics finds that urban land development restrictions in our most productive metropolitan areas are slowing the growth of the US macroeconomy.[11] If California farmers, located at the fringe of productive metropolitan areas such as San Francisco, sell their land to suburban developers and have the land rezoned for higher density, then more people could live in the productive areas and housing would be more affordable. A critic might worry that this will increase the San Francisco area's traffic congestion. Traffic congestion pricing on these major roads would mitigate this concern.

In major metropolitan areas such as San Francisco, many corporations are located in the suburbs. Some of these suburban campuses can be commuted to from what are today farming areas. High-tech workers will seek out high-quality housing within a reasonable commute time radius of their place of work. In the previous chapter, the benefits of introducing road pricing to reduce traffic congestion was discussed. If there is congestion pricing, then this commuting radius increases, and past farmland becomes extremely valuable as potential suburban housing. During the covid-19 pandemic, more workers (especially those in the tech sector) are working from home. The rise of telecommuting raises the possibility that more farmland in the vicinity of high-tech centers will be converted into housing.

If less of our food is grown in California in the future, then other land will be converted into farmland. Rising productivity per acre would mean that less land would need to be farmed to produce the same number of calories and would help offset land lost to desertification in places such as California.

Farmer Variation in Nimbleness

Farmers whose yields decline due to changing climate conditions can sell their land to other farmers with an edge at coping with change. Once these farmers have sold the land, what do they do next? The costs of farmers adapting to climate change will be higher if these former farmers have trouble transitioning into new careers and occupations. The labor economics literature yields a pessimistic finding that job-training programs for middle-aged workers have not proven to be effective at retraining.[12]

Why do farmers differ with respect to their coping ability? Two explanations include access to capital and farmer human capital. The farmland itself may need new investments to reposition it for growing new crops. Examples include drainage canals being dug, irrigation being installed, and soil being enriched differently than before. Larger farms are more likely to have the capital to finance these transitions. This suggests that small farms will be consolidated as the land is sold to larger farms. The human capital dimension is directly tied to manager quality. Higher-quality managers are better positioned to cope with change because they are more likely to be able to solve novel problems.

Some farmers may not seek solely to maximize their stream of profits. For some farmers, farming is a way of life inherited from previous generations. If such farmers have social and personal reasons to remain on their land, they hold an option to switch crops and grow varieties that can thrive under the new climate conditions. In the case of family farms, or farms operated by individuals with an attachment to a particular piece of land, these farmers will be slower to adapt to new conditions because they lack access to the same menu of adaptation strategies as a less sentimental farm. Some of these farmers are introducing new revenue streams such as farm tourism, where they welcome urbanites to visit the farm, partake in their way of life, and purchase farm-made products. Such revenue diversification helps to shield such farmers from lower output.

If farmers are unaware of the changing climate conditions, then they will be less likely to seek out new solutions. In this case, their productivity will decline as they are unprepared for changes in weather conditions. Research is needed on understanding how farmers' subjective beliefs about emerging climate risks are formed and how they change over time.

Many smaller farm operators work with extension specialists. These specialists are often connected with university agricultural programs, and they play a key middleman role connecting new ideas from the academy to farmers. Through repeat interactions, the farmers trust these specialists. Such specialists hold a unique position to identify farmers who are at risk of not preparing to adapt to changing conditions. Those farmers who know that they do not know how climate conditions are changing will be more likely to proceed with caution and to invest in growing strategies that preserve the option to reoptimize as they learn about emerging risks.

If there are competitive farming land markets, underperforming farmers can sell their land to other farmers who have the human capital and the access to finance to reposition the farmland to its new highest and best use. In this sense, agricultural farmland markets facilitate adaptation by allocating scarce land to the managers who have an edge in using this resource. No young farmer would invest scarce time and resources to become an expert in climate resilient agriculture if that farmer is unable to acquire farmland to apply the knowledge. In this sense, there are synergies between acquiring human capital and access to land markets.

The Rise of Indoor Farming

"Indoor farming" is a catchall term that refers to several types of agriculture occurring inside structures that allow for varying degrees of climate-controlled conditions for crops. The oldest and most common type of indoor farming is the greenhouse, where plants are grown inside a building with transparent walls that allow sunlight in but control for other factors like temperature and humidity. Greenhouses have been employed in colder countries, allowing for the growth of vegetables originally cultivated farther south. Greenhouses represent a middle ground between traditional agriculture and modern vertical farming techniques, using more electricity than the former and less than the latter. They also lack the total control of vertical farms, because they are dependent on the sun to provide energy to plants as opposed to human-controlled grow lights. Greenhouses also split the difference in terms of yields per unit area of land, producing more than traditional agriculture, but they are not as productive as stackable vertical farms. Labor usage tends to be higher than traditional farms, because the smaller scale and enclosed spaces of greenhouses mean that equipment such as combines or automated harvesters are either impractical or cost-inefficient to employ.

Vertical farms are reliant on artificial light to grow crops. They use more electricity than greenhouses or traditional farms. Their upfront fixed costs are higher due to the need to purchase grow lights, as well as the increased cost of the operation's building relative to greenhouses.[13] The payoff from these investments is improved performance along every other metric—water usage is significantly decreased (by about 95 percent relative to outdoor farms), pesticide usage can be all but eliminated, and yield per unit area of land is much higher as well. These gains are achieved by the different types of planting technology available as well as by the ability for plants to be grown above each other in stacked tiers (hence the term "vertical farm").

Though indoor farms are a promising technology for farmers growing high-margin crops such as spinach or tomatoes, it would be impractical for the majority of planted acreage in the United States to be replaced by an equivalent amount of indoor farming square footage. Most US non-pasture agricultural land is devoted to commodity crops such as corn, wheat, and soy. These crops are much less profitable on a per acre basis but make up the vast majority of the nation's caloric intake.[14] These commodity crops require large-scale harvesting operations that are simply not feasible in the confined spaces of vertical farms.

There are several environmental benefits if more farmers move their operations indoors. The most immediate is a significant reduction in water usage: Local Roots, a California-based indoor farming company, claims that their farms use 97 percent less water to grow lettuce than traditional methods.[15] The reduction in water usage varies across indoor farming operations, but a 90 percent reduction in water usage would significantly reduce the strain agriculture places on California's water supply. These farms could locate much closer to major population centers in California, reducing energy expenditure used to move produce to consumers, improving the quality of produce delivered to these cities, and creating more spatial diversity in California's food supply.

Not all crops can be moved indoors. Citrus crops and tree nuts are some of California's primary agricultural products, and both types of produce are hard to grow indoors efficiently due to their large size. The small size of lettuce plants allows rows of these plants to be stacked in an indoor farming operation, increasing yield per square foot of building space.

The most important change will involve how plants are watered. Relatively low-cost technologies such as drip irrigation, along with more expensive ones that actively monitor soil moisture to determine when plants need water, will allow farmers to reduce their water usage.[16] The increased need for drought-tolerant crops will create demand for biotech firms to create new strains of crops better able to withstand the heat and the likely drought conditions.

The rise of indoor farming does face policy risk. An open question is whether local governments will allow land to be rezoned for this purpose. At their core, indoor farming operations function like manufacturers: they place a high load on local utility grids, and as these operations grow to take advantage of economies of scale, they will require a constant stream of workers and transport trucks to run the operation and deliver produce. Though locating close to consumers increases the benefits of indoor farming significantly, the reality is that these operations could impose negative externalities on local neighbors. Manufacturing factories pollute the local air and water. Whether indoor farms have similar negative spillovers remains an unknown. Local governments seeking to enjoy the benefits of nearby indoor farming operations should work to develop zoning codes that can accommodate this land use. One approach local government can take would be to treat indoor farms like light industry and place them in peri-urban industrial parks. Locating indoor farms in groups on the edge of cities, near warehouses and light industry, would allow for easy transport access and limit the negative effects of these operations on neighboring communities. Furthermore, the appropriate utility infrastructure can be installed to serve the needs of indoor farms ahead of their construction, negating the need for costly utility upgrades in populated areas.

The Rise of GMOs

Scientific advances in crop science promise to improve the resilience of the agricultural sector. Large farming corporations possess both the access to capital and the scale necessary to profitably create new resilience measures for their crops. This was how genetically modified organisms (GMOs) made their way into mainstream use by farmers in the United States and then in the developing world. Such GMOs open up the possibility of reducing the climate sensitivity of agricultural output with respect to extreme climate events.

Boosted by a US Supreme Court decision allowing General Electric's scientists to patent genetically engineered bacteria, modern GMO research became a profitable pursuit for the agricultural sector to improve the quantity and quality of its output.[17] A thirty-five-year panel study on county-level corn yields in the United States finds that the adoption of genetically engineered corn in a county is associated with a 17 percent increase in yield.[18] Part of these gains in yield are due to the fact that genetically engineered crops can be made more resilient to pesticides and droughts. However, these improvements can be costly. The average genetically modified crop takes thirteen years and $130 million in investment before being brought to market.[19]

GMO-based innovation can be used to address many challenges posed by climate change.[20] GMOs improve crops' resilience to high heat and weather volatility. C4 rice is one such project, aimed at minimizing water and nutrient loss in rice plants. Because of increased evaporation due to heat and drought, water will be an increasingly valuable resource for farmers in coming years. The agricultural sector is already a major user of water, consuming about 70 percent of water used globally each year.[21] Decreasing crops' water needs will be vital not only to improving crops' resilience to drought but also to reduce the agricultural sector's water demand. A study of the effects of water scarcity in California found that water scarcity could reduce agricultural revenues by up to 85 percent in the short run; having technologies to mitigate this decline in the short run before longer-term adaptation strategies can be employed will be critical to the health of the agricultural sector.[22]

GMOs can allow crops to be grown in previously infertile locations, expanding the total area of usable farmland. Yuan Longping, a Chinese agricultural scientist, created rice that could grow in the deserts of Dubai despite the arid conditions.[23] Such innovations bring farming to previously barren lands and allow crops to be grown closer to desert-located populations hit hardest by higher temperatures. Crops can also be modified to mature faster, allowing them to be grown in regions whose seasonal growing windows were previously too short to support these crops. Thus, these genetic advancements expand farmers' options for arable land even as climate change makes large swathes of existing land unsuitable for farming.

GMOs also help crops better meet the dietary needs of the global population, improving the yield and nutrient content of crops. Vitamin deficiencies in poorer regions can lead to significant economic losses by causing debilitating illnesses. For example, a lack of vitamin A can cause childhood blindness, especially prevalent in poorer regions. To solve this, scientists created golden rice by modifying normal rice to provide half a day's worth of vitamin A in one serving.[24] As GMO technology becomes cheaper, crops can be tailored to make up for dietary shortfalls of different regions.

Though these innovations often come from global organizations or first-world firms, the price of GMO crops eventually decreases enough to be marketed to farmers in the developing world. The additional demand from nonadvanced countries creates a wider market for GMO developers, encouraging them to put more money into research. Climate change is expected to be the most disruptive in Africa, where the agricultural sector is dominant.[25] Because of this, global demand for GMO crops can be expected to increase significantly as climate change worsens, encouraging further GMO research.

The adoption of GMO technology does pose new ambiguous risks.[26] Though designed to affect only a specific gene, gene-editing techniques can have unintended consequences for other aspects of a crop's genetic makeup.[27] Without the proper testing processes in place, unexpected vulnerabilities in modified crops could arise. Furthermore, certain modifications, such as pesticide resilience, may encourage greater use of pesticides and harm the local ecosystem.

In this big data age, governments and farmers can collaborate to run a randomized field experiment to deploy GMO technology to learn more about its promise and pitfalls. The 2019 Nobel Prize in Economics was awarded to three scholars who implement such field experiments in the developing world. Their focus is to identify cost-effective strategies to reduce poverty and improve the quality of life of the poor. In a similar spirit, new field experiments could be launched in US agriculture to measure the intended and unintended consequences of deploying GMO technology. A key point emphasized by ongoing economic research is that there are heterogeneous treatment effects. In plain English, this means that the same intervention (such as adopting a GMO technology) will have a different impact on different farms. This diversity in outcomes is likely to be due to both the soil conditions and the farmer's ability to optimize working with the new technology. GMOs will offer greater benefits if farmers have the human capital and the financial capital to achieve the best results. Larger farms managed by high-quality human capital managers will be more likely to gain the most from this technology. This logic suggests that such new technologies are likely to increase earnings inequality among different farms, but the overall food supply purchased by urbanites will be increasingly resilient.

The research documenting the sensitivity of conventional yields to climate conditions highlights the importance of encouraging experimentation to learn about the costs and benefits of emerging technologies. If agricultural producers and regulators acknowledge that we know that we do not know the full consequences of GMO adoption, then we can launch the equivalent of clinical trials (borrowed from the drug testing literature) on different plots of land that face different climate conditions to learn about the net benefits of GMOs. Indoor farming will allow even more experimentation as firms could replicate a variety of climate conditions within the same facility, allowing for more rapid development of GMO strains suited to particular climate conditions.

In recent decades, much crop adaptation has largely been reflected only in improved yields, while drought and heat resistance has not improved much (some crops have even become more vulnerable to drought conditions).[28] As the probability of drought events rises, farmers who recognize this trend will seek out growing strategies that are more resilient in the face of drought. Such aggregate demand creates opportunities for those who develop GMO varieties that perform well under the new climate conditions.

Crop resilience does not end at the harvest. The transportation of crops can be costly, especially for delicate produce that requires specific humidity and temperature conditions to stay fresh. Spoiled crops mean lost profits. Crops that can maintain their quality through long trips lower food waste but also make it easier to import them from distant regions. This will be a key adaptation pathway if climate change prevents some crops from being grown in certain regions.

Big Data Brought to the Farm

Before the age of big data, farmers relied on their memory of historical climate variability. Such information helped them to adapt to growing conditions for a given crop in a specific region. Though weather conditions might deviate year to year, farmers could expect a relatively consistent climate. Farms would pick crops and farming techniques best suited to their land, based on historical averages of temperature and rainfall. In a stable climate, they could expect their local knowledge to have future predictive power.

Climate change promises to turn this dynamic on its head. Changes in long-term weather trends will require the agricultural sector to quickly recognize and respond to conditions it has never faced before. Due to climate change, farms can no longer just assume that future weather trends will mirror past trends. Bad weather one year may be more than just a one-off event but instead reflective of a long-term trend. The agricultural sector will become increasingly reliant on data and statistics to spot these changes early on.

The use of big data helps farmers to cope with novel growing conditions. A statistician can now access millions of records concerning different farmers' production of different agricultural outputs and data on the farmers' production inputs (labor, land, nutrients) and weather conditions. Machine learning techniques allow the data scientist quickly to figure out which climate conditions favor growing which crop.

Consulting firms will produce such reports and then sell this information to farmers. This information will boost farmer productivity. The machine learning approach can provide exact information for the farmer concerning water inputs and other approaches to calibrate the key inputs in production. A combination of improved sensors and increasing automation makes this a reality for the agricultural sector.

Farmers have embraced the Information Age, piggybacking on technological developments that on their face have little to do with farming. Farmers are using drones as a low-cost way to monitor their crops visually and to inspect their land for water leaks in irrigation systems.[29] The Internet of Things, aimed at putting sensors in everyday objects, is still in its infancy, but farmers are already equipping their fields with sensors to collect data on a massive scale. Sensors can be placed throughout a farmer's land to monitor soil quality, moisture, and even crop ripeness.[30] Farmers save the time of manually collecting this data and can collect information on a much larger scale than previously possible.

For example, the decision agriculture firm Arable Labs creates mounted sensors that assess conditions both above the soil and within it.[31] Metrics include light levels, soil moisture, precipitation, and evaporation. Collecting data every five minutes and transmitting it hourly or daily, the sensors aggregate data for farmers and allow them to respond in near-real time. Farmers are saved the effort of manually checking these conditions and can direct additional resources to areas that need them or avoid wasted resources, such as overwatering. Using the data collected from the sensors, machine learning statistical tools are used to generate short-term predictions of weather conditions.

Farming equipment is making the transition away from human operators, borrowing technology from the development of driverless cars. The benefits of automated farming machinery are many. Climate change is expected to drive average temperatures upward, potentially lowering the productivity of farmers and limiting their outdoor working hours. Machinery, in contrast, can be more easily insulated from the effects of high outside temperatures. Furthermore, automated farming equipment can skip the human intermediary, directly interpreting data from crops and quickly administering the optimal amounts of water and fertilizer to each plant.

Big data can also benefit the agricultural sector more widely, accurately predicting weather impacts on crops on a regional scale and giving farmers time to protect their land. These technological solutions hold promise for boosting the resilience of traditional farmers in developing nations. The African Risk Capacity, an agency of the African Union, uses satellite weather surveillance software to predict drought risks to poor farmers during the growing season long before losses are recognized during harvests.[32]

The benefits of predictive software extend to farmers in advanced nations. In response to early warnings of drought, farmers can spray their crops with chemical compounds to protect them against environmental stresses such as high heat.[33] While these measures are expensive to use every growing season, they can be a vital last-ditch measure to avoid total crop failure. Using satellite data to farm more efficiently would be far too expensive for a single farmer to undertake, but the average cost per farmer is much lower when such information is shared throughout a vulnerable region.[34] Information is a local public good. The productivity of a region's agricultural sector is enhanced if the nation has invested in a satellite program for engaging in such remote sensing. A team of researchers at the

University of Illinois created an agricultural yield model that incorporates satellite data along with seasonal climate predictions. This model's predictive success highlights the role that satellite data will play in helping farmers in the future.[35]

Armed with big data, smart farms can also better perform their own farming field experiments. Software programs can be designed to identify long-term trends in local growing conditions. Farmers can use this information to decide whether to continue farming the same crops or transition to new crops or growing techniques. Predictions of soil degradation based on in-ground sensors can encourage farmers to take preemptive actions like fertilizing the soil to prevent their land from becoming infertile. Struggling farms can also better recognize when they are fighting a losing battle against worsening growing conditions and cut their losses earlier by selling their land. Recently, researchers have also used geographic information systems data to identify unused areas of land that are profitable to convert to farmland while marking other areas to be protected from deforestation.[36]

Real-time data on the output of different farmers can be used to study how both policy shifts and weather shocks affect their productivity. The Trump administration's trade war with China, and China's retaliatory tariffs on US soybean exports, has created a natural experiment to study how farmers respond to changes in the rules of the game. This international relations turmoil has meant that American farmers face a much lower soybean export price.[37] In recent years, many US farmers had switched from growing corn to growing soybeans, due in large part to rising Chinese demand for soybeans. After the imposition of the Trump tariffs, many farmers switched back to corn. This case study highlights that many farmers in the United States are close to the margin for growing one crop over another. Using farmer-level data on the choices they make and how these choices vary over time will provide key insights concerning what fraction of farmers are nimble. Are the nimblest farmers acquiring more farmland over time? Is the net result of this market transition a more resilient agricultural sector such that the correlation between agricultural output and extreme heat is shrinking to zero? Such evidence would support the claim that the agricultural sector's adaptation progress is accelerating.

Educating the New Generation of Farmers

Education will play a vital role in accelerating adaptation progress. Spreading knowledge of new farming techniques has already had significant impacts throughout the developing world. One study finds that farmer education in Nigeria significantly increases productivity when coupled with technological innovations and encourages further adoption of new technology.[38]

Wheat, corn, and rice are staples in the world's diet, providing more than 42 percent of the calories consumed globally.[39] But climate change may necessitate the rise of more resilient crops as new global staples. Farmers can learn to grow such crops as wild grasses, perennials that can survive with less water and grow year-round. The growing patterns of such perennial crops require different cultivation methods than many farmers are accustomed to, meaning that farmers will need a way of acquiring the knowledge necessary to switch crops.

Small farmers cannot afford to leave their lands for months or years to learn the latest farming techniques. Instead, agricultural experts have had success by directly reaching out to farmers at a local scale and teaching farming communities techniques to maximize their productivity and improve their resilience to extreme weather. Farming field schools set up in Southeast Asia follow a decentralized approach, providing weekly, hands-on farming education to groups of around two dozen farmers at a time.[40]

In Southern Africa and other locations around the world, the Farmer-to-Farmer Program, funded by the US Agency for International Development, operates in a similar way.[41] American volunteers travel directly to poor farming communities to provide technical expertise for improving agricultural productivity. This program is a prime example of closing the knowledge gap in the agricultural sector, ensuring that farmers are using the most effective techniques available. As innovation increases in response to the pressure of climate change, it will be especially important to ensure that this new knowledge is adopted on a wide scale through education. Apprenticeships, associate degrees, and even bachelor's degrees can connect farmers with the latest tools dreamed up by agricultural scientists.

The importance of educating farmers will become especially vital as climate change transforms the environment farmers are used to working in. Higher temperatures and more extreme weather require farmers to adopt farming techniques they did not need in the more stable climate of the past. Farmers may know how to protect their crops against the occasional drought or storm, but they will need to learn new approaches when these conditions become the norm, rather than anomalies.

Access to Financial Markets

The agriculture sector's adaptation efforts will require investing new capital. The financial sector supplies this capital. This means that, in the United States, the interactions between Wall Street and farmers will play a key role in determining the timing and the extent of the transition. Farming in the United States is a capital-intensive industry, with debt levels reaching $395 billion in 2016 and continuing to rise.[42] Because of this heavy reliance on private capital, farmers will struggle to operate if financiers deem them unworthy of loans. Lenders have an incentive to avoid loaning to firms that underperform currently or fail to protect themselves against climate risk. Financial firms that loan money to farmers have accumulated experience in this sector and have the expertise to judge which farmers face new climate risks. Such investors, similar to the insurers and mortgage lenders, are the adults in the room. In the case of farming, lenders can play a similar role nudging farmers to invest in resilient strategies. If such farmers realize that they face more stringent borrowing terms because they are not coping with climate risks, thus becoming less profitable and in greater danger of defaulting on their loan, then this nudges such farmers to take new steps to cope.

Futures markets allow investors to make bets on the direction of prices for corn, wheat, soy, and other commodity crops. Futures prices aggregate the knowledge and expectations of various investors, farmers, and other financial actors, creating signals for the predicted performance of these crops. Farmers can use trends in futures prices to predict the price they will receive when they bring their harvests to market. There are many futures markets that do not exist. Today, there is no futures market for selling oranges fifteen years from now. Futures markets do not extend that far into the future. If such futures markets did exist, then the prices of these contracts would convey important information for farmers who are making medium-term plans. If the fifteen-year futures contract on oranges features a high price relative to today's price for oranges, then this signals that investors are pessimistic that future supply will keep up with future demand. This signal would have beneficial adaptation effects as suppliers who feel they can grow oranges in the face of emerging climate patterns would invest more effort to grow more oranges.

Climate change affects regional and global weather trends, making it hard for insurers to diversify the risks crop insurance covers. A disaster such as a drought can affect the farmers of an entire country. As a result, an insurer would face numerous claims at once, threatening its ability to make payouts. Insurers that cannot diversify against this risk will raise their premium prices or simply refuse to offer coverage. Alternatively, insurers can become more creative with the types of insurance products they offer to farmers. This insulates both farmer and insurer from the costs of climate change.

One example of how access to finance and specialized human capital work together to build up resilience is African Risk Capacity. The program pairs its weather-monitoring satellite software with risk pools made up of several African countries. Member states pay premiums for varying levels of insurance coverage, much like insurance marketed to consumers. Since participating states are located across the continent, the program diversifies against the risk that a weather event strikes all nations at once. The program operates via parametric triggers, payments from the insurer to the owner of the insurance policy when a specific climate event occurs (such as extreme drought), rather than when the actual insured asset is damaged. The contracts help the insured to reduce the financial impact of a failed harvest triggered by outlier climate conditions.[43] Participating countries also have access to the Extreme Climate Facility, a financial mechanism that triggers extra, private funding to these states if extreme weather ensues.

In the near future, more private insurers will likely sell parametric insurance contracts (tied to actual climate events) as more comparable climate sensors are installed across different geographic areas. Insurers will be able to refer to a land parcel's sensor data to determine whether a claim warrants a payout. Such sensor data also allow insurers to judge the severity of weather events more easily and offer different payout levels accordingly. If the land parcel sensor data are in the public domain, then competition among insurers for attracting customers will mean that insurance buyers will have more choice over what types of insurance contracts they can buy, and the contracts will be cheaper to purchase. The ability to purchase such an insurance contract means that farmers will have peace of mind knowing that they will receive money from the insurers if bad climate events occur. This reduces their family's consumption risk exposure and lessens the economic damage caused by climate change. This risk smoothing (for example, that the farmer's income will never fall below a threshold even if the weather is awful) is especially valuable to risk-averse farmers.

Financial firms also play a key role in the funding of new agricultural innovations. Agritech is still a nascent industry, and as a result funding from many mainstream investors such as pension funds remains sporadic due to the perceived risks of the agricultural start-up industry.[44] Even so, venture capital in agriculture continues to rise at a remarkable rate, climbing ten times higher than 2012 levels to more than $4.6 billion in 2015.[45] Venture capital can fill the gap in funding for agritech firms developing promising technologies but lacking the size and consistent revenue streams necessary to secure funding from traditional sources like banks.

Adaptation-Friendly Public Policies for the Farming Sector

Government agricultural policies greatly influence this sector's investment choices. Farming is one of the most heavily subsidized sectors in the United States, with the federal government paying out $17.2 billion in 2016.[46] These subsidies create significant distortions in the agricultural market. Farmers exert outsized political influence and have secured favorable government policies. Although farmers benefit greatly from these policies, the costs are distributed among taxpayers. Because each taxpayer faces only a small fraction of the total cost, inefficient policies largely go unnoticed.

Government policies can provide a vital boost to farmers' adaptation efforts, or they can entrench the status quo to the detriment of farmers and taxpayers. A pro-adaptation agricultural policy agenda would foster basic research on agricultural output resilience and would encourage the transfer of scarce resources (such as land) to those with the human and financial capital to make the best use of it.

Farmers rely on federal crop insurance to reduce their risk exposure. The vast majority of commodity crops grown in the United States are insured, including about 90 percent of corn and soybeans.[47] In 2017, farmers faced an average deductible of 26 percent on their crop insurance claims.[48] While these percentage deductibles are steep, the alternative of going uninsured can be enough to put a farmer out of business if a severe disaster strikes. Thus, there is a strong incentive behind the high participation rate in these crop insurance programs. However, these high deductibles mean that farmers may be strapped for cash at the very times they realize they need to invest in greater resilience measures. One way to encourage farmers to adapt would be to reduce deductibles for specific adaptive improvements, giving them an immediate incentive to invest in protecting their crops.

A harsher climate threatens farmers' profits, and they are likely to seek out more government support. Agribusiness spent over $130 million in 2017 to influence policy makers, more than defense lobbyists spent in the same year.[49] The fact that each state has two senators guarantees that farm states (while not heavily populated) will continue to have strong political influence. While policies such as government-subsidized crop insurance are cheaper for society when the climate is stable, a harsher climate means that these policies can become significantly more costly to government and ultimately to taxpayers.

Subsidies to the agricultural sector have gone under the radar for decades, largely because they make up just under 1 percent of the government's annual budget. Increased farming losses will drive this percentage upward.[50] These distributed costs of farm subsidies have been too small to warrant voters pursuing information on how exactly their taxes are being spent. This barrier enables pork barrel politics, in which politicians grant benefits to their local constituents funded by national tax revenues.

In recent years, the Internet has increased government transparency through the creation of such databases as OpenSecrets, which publishes detailed statistics on elected officials receiving campaign contributions from different sources. This increased ability for the media to follow the money in politics fosters accountability. The Nobel Laureate Gary Becker argued that, as the inefficiency of government subsidies rises over time, so political reform becomes more likely.[51] Climate change, through increasing the cost of government agricultural subsidies, will offer an important test of this hypothesis.

Current US farming subsidies disincentivize resilience and efficiency in the agricultural sector.[52] Such public crop insurance, combined with subsidized deductibles, insulates farmers from swings in output prices and yields. Suppose a drought strikes the Midwest and drastically reduces crop output. Under the current subsidy program, insured farmers would receive payments from the government and can even make a profit on that harvest. If farmers make a profit even when a drought hits, why would they spend money protecting their crops against future droughts? Why would these farmers consider switching crops to grow a new crop that faces less risk from the new climate conditions? Adverse conditions drive innovation, but only if the decision makers face severe consequences for failing to adapt.

Crop insurance and commodity crop support programs, both of which disincentivize innovation by farmers, together make up over three-quarters of the farm subsidy budget. Policies such as crop insurance insulate farmers from losses due to bad weather, but these costs are paid by the taxpayers. If the farmers face no repercussions for a bad harvest, they will simply continue farming as they have always done, and taxpayers will continue to foot the bill.

State and local policy that subsidizes natural resources such as water also disincentivizes farmer adaptation. In California, farmers face extremely low water prices per gallon relative to what urbanites pay. This implicit subsidy causes some farmers to grow water-intensive crops such as alfalfa when they would make a different decision if they faced the market price of water. If these farmers had the property right to sell the water that they do not use, then such farmers would grow less alfalfa and sell more water to urbanites. This example highlights how subsidies muffle the scarcity signal that resource prices are meant to send. As climate change increases California drought risk, this issue will only become more important over time.

Complementarities between Private and Public Investment in Basic Research

Government can take a direct role influencing the intensity and direction of agricultural innovation. Public-private partnerships between scientists employed by the government and large farming firms already do this. On their own, private researchers tend to direct their efforts toward advances that will directly produce profits for their employers. For example, a private researcher in the United States may develop a gene insertion technique to improve a crop's yield and pesticide resistance. Qualities like vitamin content mean little to the researcher, because the vast majority of American consumers do not suffer from nutrient deficiencies. Public researchers often have greater freedom to pursue innovations with profit outcomes that are either less certain or more long-term in nature, both being key challenges in the fight against climate change.

The National Science Foundation pursues a competitive grants program that helps to build up basic science. Such grants provide academics with the financial resources to purchase specialized equipment and to release them from teaching to have the time to conduct research that pushes the research frontier forward. To offer just one example, Joyce Van Eck and her colleagues propose using the genome editing tool CRISPR to increase crop productivity of producing groundcherry (a cousin of the tomato). This fruit offers high nutritional value; it is packed with vitamins C and B, beta-carotene, and antioxidants.[53] Once a science team such as Van Eck's perfects a new variety of agriculture, such blueprints can be widely used by farmers to introduce this variety. This trickle down from bench science into the field offers adaptation benefits because the knowledge is in the public domain. The Nobel Laureate Paul Romer would emphasize the increasing returns to scale here. Once we have the blueprint, the average cost to society of adopting this new variety declines as the number of users increases.

Both NSF-funded academic research and government-employed researchers enhance our stock of public farming knowledge. As these researchers make progress on core farm problems such as pest control, soil quality maintenance, and seed quality and resilience, these breakthroughs raise the productivity of individual farmers. As climate change shifts climate conditions, more of these researchers will focus on climate challenges, and this competition between scientists will yield valuable breakthroughs. This pace of progress will be accelerated if the NSF has a larger research budget. An attractive feature of the National Science Foundation is its commitment to peer review and avoidance of picking winners. The NSF takes pride in having open competitions for research funding such that each proposal is peer reviewed by independent experts who do not have a conflict of interest. Their rankings of proposals are submitted to another panel of experts who convene to determine which proposals will receive funding. The ranking of proposals is based on their scientific merit, the quality of the investigators, and their core research goal. Together this process means that American taxpayers receive a high bang per buck spent on basic research.

This chapter has covered many adaptation margins that farmers will have access to. This large menu creates a certain degree of optimism about the future of farming in the face of climate change. Given that the world's population is growing and must eat to survive, the anticipation that the food supply is threatened by climate change creates a tremendous profit opportunity for those future farmers who are nimble.

Every day, we learn new lessons about reducing our risk exposure. A silver lining of the covid-19 pandemic is that farmers have learned valuable lessons about maintaining food supply chains during a crisis. The media have reported that at the same time that some poor urbanites are hungry, food is rotting on the farm. More resilient food supply chains will emerge as farmers seek to diversify the buyers for their products. Both food processors and supermarkets will also seek to diversify their sources for supplies.

11

Globalization and International Trade Facilitate Adaptation

Globalization, defined as low barriers to international trade, increases one's potential trading partners. If a person lives in a nation of ten people and this nation is closed to international trade, then this person can only trade with nine other people. The possibility of trading with everyone on the planet opens up a huge menu of possible alternatives.

Removing barriers to international trade in labor, capital, and goods facilitates climate adaptation. As the shocks experienced by specific places intensify, the ability to migrate farther distances and to import goods and capital helps people to cope with local shocks. This benefit of globalization is especially important to consider given the Trump administration's policies that have limited trade and labor flows between the United States and other nations such as China.

In 2020, efforts to contain the novel coronavirus are ongoing. Global air travel accelerated the spread of this disease, and this highlights how globalization imposes costs. Measuring the costs of globalization is an important task because these costs are not equally distributed. Globalization exposes incumbents to more competition because buyers now have more choice in purchasing goods and in borrowing money. In the United States, as some factories have closed and others have moved abroad, richer people have faced less earnings and unemployment risk than poorer people.

Although globalization imposes costs, most economists believe that it offers even greater benefits, and these benefits are likely to rise over time as we face more climate change risk. By being able to trade with others, people can focus on developing specialized skills, collect income from this talent, and then use markets to purchase other goods and services that they want to consume. When trading possibilities shrink (because of immigration walls and tariffs), this has the net effect of lowering incomes and well-being.

International Migration and Climate Change

Limiting international migration significantly lowers the quality of life of individuals seeking to adapt to shocks at their origin. Research studying the effects of the Dust Bowl in the 1920s documents that migrants' quality of life improved as they moved away.[1] Migration has two important effects. For those who choose to leave a place that has experienced a recent shock, these migrants will seek out the right place for them. As they move, they are likely to earn more income in their new destinations, and this will help their families. Economists stress a second effect. If enough people leave the area that experienced the bad recent shock, then this actually increases the quality of life for those who remain in the afflicted area. For those who remain, the logic of supply and demand predicts that their rents decline and their wages rise as people move away. If people are leaving a rural area, then this urban migration means that there is more land per remaining rural person, and this will help to reduce their poverty risk.

Climate scientists have posited that future drought and heat in the developing world will create large numbers of climate refugees. As crops fail and livestock die, desperate people will resort to any strategy to survive. Migration across international political boundaries can help people to avoid these risks. While many poor people in developing nations want to move to richer nations, the richer nations have the property rights to determine who is allowed to enter.

The current backlash against international migration inhibits the ability of people in developing countries from moving to Europe and other richer areas. In smaller nations with fewer cities, rural people will have fewer domestic destination choices. People from these nations who confront new climate stresses might seek to migrate but face limited domestic options. If the option to migrate across nations continues to be available, then poor people in the developing world will have a much better chance to adapt to emerging threats facing their location of origin.

Migrants consider what their quality of life would be like in each possible destination. This requires imagination about what their life would be like in a place they are likely never to have visited. Migrants are less likely to move greater distances. Controlling for the distance from one's origin to a possible destination, migrants are more likely to migrate to an area where their skills are in demand and where the local culture and language welcome them. Migration destinations face potentially large short-run costs when large numbers of migrants enter. First, there is the short-term increase in the cost of providing services for the suffering migrants. Such individuals need health care, housing, and education. Incumbent low-skilled workers now face more competition as the new immigrants vie with them for jobs. From the basic logic of supply and demand, a nation experiencing an influx of migrants will experience a declining wage for low-skill work and rising housing rents.[2] Second, issues of cultural mismatch and assimilation arise. In a world featuring different religions and cultures and past historical bad blood, incorporating hundreds of thousands of newcomers raises key issues in preserving the social fabric and character of the destination area.

Even though immigration imposes short-run costs on the destination nations, these nations benefit as they experience an influx of younger workers. Researchers at the World Bank estimate that "the average income gain for a young unskilled worker moving to the United States is estimated to be about $14,000 per year. If we were to double the number of immigrants in high-income countries by moving 100 million young people from developing countries, the annual income gain would be $1.4 trillion."[3]

Many European nations and the United States feature an aging population with large promised entitlements for Social Security and pensions. Pay-as-you-go retirement schemes are much more likely to be financially viable if the nation features a high ratio of young workers relative to the count of retirees. Both the United States and European Union feature large ratios of elderly relative to younger people. These nations face the prospect of a huge tax burden being imposed on the young to finance the streams of benefits promised to the elderly.[4]

Economists admire the invisible hand of capitalism. Migrants seeking a new and better life have strong incentives to seek out new destinations offering high wages. The high wage acts as a scarcity signal indicating that the destination area needs young talent. In this sense, the market price system acts to direct the migrants to the right destinations.

Immigration Market Design

Back in 2005, the Nobel Laureate Gary Becker proposed that a new market in passports would allow immigrants to buy their way into moving to another country:

That is why I believe countries should sell the right to immigrate, especially the United States that has so many persons waiting to immigrate. To illustrate how a price system would work, suppose the United States charges $50,000 for the right to immigrate, and agrees to accept all applicants willing to pay that price, subject to a few important qualifications. These qualifications would require that those accepted not have any serious diseases, or terrorist backgrounds, or criminal records. Immigrants who are willing to pay a sizeable entrance fee would automatically have various characteristics that countries seek in their entrants, without special programs, point systems, or lengthy hearings. They would be younger since young adults would gain more from migrating because they would receive higher earnings over a relatively long time period. Skilled persons would generally be more willing to pay high entrance fees since they would increase their earnings more than unskilled immigrants would. More ambitious and hard working individuals would also be more eager to pay since the U.S. provides better opportunities than most other countries for these types. Persons more committed to staying in the United States would also be more likely to pay since individuals who expect to return home after a few years would not be willing to pay a significant fee. Committed immigrants invest more in learning English, and American mores and customs, and become better-informed and more active citizens. For obvious reasons, political refugees and those persecuted in their own countries would be willing to pay a sizeable fee to gain admission to a free nation. So a fee system would automatically avoid time-consuming hearings about whether they are really in physical danger if they were forced to return home. The payback period for most immigrants of a $50,000 or higher entrance fee would generally be short—less than the usual pay-back period of a typical university education. For example, if skilled individuals could earn $10 an hour in a country like India or China, and $40 an hour in the United States, by moving they would gain $60,000 a year (before taxes and assuming 2000 hours of work per year). The higher earnings from immigrating would cover a fee of $50,000 in about a year! It would take not much more than four years to earn this fee even for an unskilled person who earns $1 an hour in his native country, and could earn $8 an hour in the U.S. These calculations might only indicate that $50,000 is too low an entrance price, and that an appropriate fee would be considerably higher. But with any significant fee, most potential immigrants would have great difficulty paying it from their own resources. An attractive way to overcome these difficulties would be to adopt a loan program to suit the needs of immigrants who have to finance entry.[5]

Imagine if Becker's rules were adopted. In a world facing climate change risk, families in the developing world would invest in the skills of one of their young people to prepare to move to the United States. By the definition of a market clearing price, there would be no rationing of slots to move to the United States. Those who could pay the fee would know they could move.[6]

In preparing to move, young people would receive an education to develop their skills to help them thrive in the United States. Families in the origin nation would recognize the economic payoff of achieving this goal.

The family preparing to send a young person would pursue this strategy because they would anticipate that this young person would send remittances back home. Research set in rural India has documented that parents invest more in their children's skills when there are greater urban labor market opportunities available to them.[7] Given the returns to skill in developed nations, these young people will earn more money and will be able to send remittances home. The families who receive these cash transfers will be able to spend more money on self-protection products such as air conditioning and higher-quality food and housing, and this will reduce this family's climate risk exposure.

Could this international market for talent create a brain drain for the origin nation as the nation's finite supply of highly skilled, motivated people leave? For smaller nations, in the short run this is certainly possible. In the medium term, though, the expectation that well-trained individuals will have a higher probability of migrating abroad to earn higher returns on these skills will stimulate several supply-side responses. Parents will invest more in their children's development if they expect that the children can migrate to the United States. Such parents will seek out elected officials and local schools that support policies to promote these skills. Together these individual choices accelerate the growth of the nation's stock of human capital. If more children seek a better education in these developing countries, then this will create an incentive for the government to offer higher-quality education.

In 2018, Eric Posner and Glen Weyl published their book *Radical Markets*. One of their novel policy proposals is the idea of allowing each person in the United States to issue one international visa.[8] In Becker's proposal, he does not explicitly state what the United States and other destination nations should do with the passport revenue they collect. This could add up to a huge amount of money. If a million people a year migrate to the United States (a nation featuring 330 million people) and each pays $50,000, then this would equal $50 billion a year in new revenue.

Given that some natives will be worse off due to immigration owing to increased competition for jobs and housing, it is crucial to deal with the losers to minimize a backlash against international migration. Posner and Weyl's proposal addresses this concern. Each native in a desirable country would search for an international migrant with whom she or he can work on a project. Posner and Weyl explain how such a contract could benefit both parties. As they write, "Immigration expands the economic pie but gives too meager a slice to ordinary people."

Thus far I have discussed an orderly migration process and the role that markets and new rules can play. During a refugee crisis triggered by war or other shocks, events quickly change as large numbers of people seek to migrate. Those nations that seek a younger workforce will be more willing to welcome migrants than another nation worried about the impact of immigrants on natives' quality of life. For those nations willing to welcome refugees, will the rest of the world offer these good Samaritans compensation for opening their border? If more nations are willing to welcome such migrants, this increases the menu of possible destinations, and this improves the expected quality of life for the refugees.[9] This logic suggests that a mechanism such that one nation compensates another nation for welcoming more immigrants would accelerate risk mitigation.

Once a nation has accepted immigrants, a new issue arises concerning how the newcomers assimilate into the nation. The new migrants know their skills and talents, and they will move to the best city for them. In the United States, many immigrants move to cities where previous immigrants from their nation have already moved. One example is Utica in Upstate New York.[10] As immigrants move to low-rent cities, this increases local housing demand and helps to stabilize home prices. This example highlights how immigrants can play a productive role in boosting the local economy in declining cities. In chapter 3, I discussed the poverty magnet effect in Detroit that has played out over the past few decades. During the auto production boom of the 1950s, Detroit's home builders built highly durable homes. When the factories closed in recent decades, the durable homes remained and became extremely cheap as people left Detroit. In the year 2020, Detroit has a large immigrant population. Such immigrants help to stabilize a city by adding upwardly mobile, ambitious young adults to an area that might have continued to decline had they not arrived. An open urban economics question focuses on the impact of such immigrants on the quality of public schools and civic life. Do the new immigrants assimilate into their new cities, or do they self-segregate? The answers to these questions will vary from case to case, but they will determine how a city's overall quality of life is affected by a surge in immigration.

International Trade in Goods and Agricultural Output

In 2018, the Trump administration raised tariffs on many Chinese imports. If other nations raise trade barriers on products such as food, then this hurts the urban poor's ability to cope with place-based risks. When small nations remove tariffs and quotas on imports of foreign agricultural produce, they encourage trade among the nation's urban consumers of food and foreign producers. Free trade among nations insulates urbanites from local climate shocks to agriculture.[11] This means that food prices at the supermarket do not fluctuate much even if the nation's farmers experience heat and drought. At any point in time, there are other food producers in other nations who are enjoying a good harvest, and the farmers in these nations can export goods.[12]

The converse also holds. If a nation closes itself off from international trade in agricultural goods, then urbanites can only eat food grown in that nation. If extreme weather conditions limit production, then local food prices will rise. Although such a price increase does not really lower the purchasing power of the rich (because they spend only a small percentage of their income on food), it does significantly lower the purchasing power of the poor.

Development economists study the political economy of whether government favors domestic farming interests over urban consumers.[13] As nations urbanize and thus more voters live in cities, this creates a political constituency that favors international trade in agriculture. As climate change creates more food price volatility in nations that are closed to international trade, it remains an open question whether elected officials will reduce these trade barriers. If these trade barriers decline, then the urban poor will face less climate risk because their access to nutrition will be maintained even when domestic climate shocks occur.

International trade in agriculture breaks the link between what a nation produces and what it consumes. Such trade acts as an implicit insurance policy against domestic climate shocks. If a nation experiences severe drought, its urban consumers can continue to eat fruit if the nation imports such goods. This discussion highlights the key role that international trade policy will play in helping consumers to adapt to climate change.

International trade is facilitated by larger transportation ships and improved storage techniques that have greatly lowered the costs of moving food around the world. The rise of ultra-large container ships has created new economies of scale for maritime trade. According to a report by the Organisation for Economic Co-operation and Development, the maximum size of container ships doubled over the decade ending in 2015. This size increase decreased total vessel costs per container by roughly one-third. Though the economy-of-scale benefits of these megaships may be reaching an upper limit, other technologies have the potential to decrease shipping costs still further.[14]

Nations that have the transportation infrastructure to move such goods from ports to final consumers will be better able to insulate consumers of agricultural goods from local climate shocks. A research team at the Massachusetts Institute of Technology developed a new form of sensor capable of detecting ethylene gas, a byproduct of ripening fruit. Current implementation of ethylene gas sensors is limited to large warehouses due to the cost of the systems, but the researchers' carbon nanotube-based system could potentially be produced at a cost below a dollar per sensor—cheap enough to be used on individual packages of fruit during shipment.[15] A newfound ability to manage fruit ripeness while in transit has the potential to drastically reduce the amount of produce spoiled in transit. It will also enable fruit to be shipped over longer distances, limiting the effects of localized climate shocks to the global food supply chain.

Free trade goes beyond trade in agricultural products. Free trade in durable products ranging from solar panels to air conditioners to refrigerators and drones reduces the prices that consumers pay for goods. Global supply chains allow nations to focus on their comparative advantage, and this specialization allows consumers to pay lower prices for higher-quality goods. The menu of possible adaptation strategies is enhanced when consumers can purchase goods at lower prices. If a nation protects domestic producers by enacting trade barriers, then consumers will have less choice and pay higher prices. If entrepreneurs anticipate that free trade will continue, then they have larger incentives to design products that they can sell on the world market. Imagine an extreme case in which Russian inventors can sell their products only in Russia. This may not be a sufficiently large enough market to bear the risk and upfront costs of designing a new adaptation product. If these inventors knew that they could sell the product on the world market, then they would be more likely to pay the upfront costs in terms of time, money, and effort. In this sense, the expectation of continued access to international export markets stimulates entrepreneurial effort.

International Trade in Capital

In a global market for capital, nations and cities that are perceived to be good credit risks will borrow at a lower interest rate. Investors are always looking for projects to invest in featuring a high rate of return and low risk. Throughout this book, I have discussed different public sector resilience projects (sea walls) and private sector resilience projects (high-rise buildings) that require capital to build. An international capital market facilitates the construction of these projects.

In a world economy featuring significant barriers to capital flows, developing nations will borrow at a higher interest rate because of their lack of internal capital. Climate adaptation projects can be capital intensive. Such projects range from paying for coastal defenses to elevating airports to upgrading sewer infrastructure to prepare for heavy rainfall. To finance these expenditures, developing nations will seek access to capital. Developing nations compete to attract international capital from richer nations. Those developing nations that develop a reputation for using foreign capital effectively (and paying back their debts) will be able to borrow at a lower rate of interest. By being able to borrow at such a lower interest rate, these nations will be able to implement more adaptation-friendly projects and will face less future climate risk.

In addition to helping people, firms, and governments to invest in ex ante resilience, access to such finance reduces consumption risk after a natural disaster. Following a major disaster, people can starve if they have no access to capital to finance their current consumption. Markets for catastrophe bonds allow affected places that have purchased such bonds to receive a financial payout that helps the area to rebuild after a disaster. In this case, the private sector is fueling the recovery.

In the Philippines, the government has been issuing catastrophe bonds. These bonds guarantee that if this nation suffers from a verifiable natural disaster, this triggers a payment by the bondholders to this nation. This guarantees that the nation has access to private capital just when its people greatly need such investment.[16] Such access to private international capital markets acts as a substitute for relying on the federal government for transfers. It remains an open question what a local government that receives such a cash infusion actually does with the money. If the money is productively spent on new infrastructure and safety precautions, then the area will be less likely to become a poverty magnet.

The globalization of the insurance industry allows private insurers to diversify spatial risk. To appreciate why this is important, consider an extreme case in which an insurer can sell policies only in Florida. This insurer holds a very undiversified portfolio. If no major storms hit Florida, then the insurer earns large profits because the entity collects plenty of revenue from policies sold and faces no insurance claims. But if a disaster hits Florida, this insurer will lose money and may go bankrupt. In this case, those who own policies will not receive a payout just when they need it. The globalization of the insurance industry guarantees that insurers do not face this risk because at one time most of the world is not hit with a disaster. By collecting revenue on policies sold all over the word, the global insurer has diversified its revenue stream and thus has the financial capacity to pay out to recent victims.

The climate change adaptation benefits from encouraging globalization are many. These points stand in contrast to the emerging nationalism we see in nations around the world. The rich gain greatly from globalization. For the leaders of democratic nations to continue to support globalization will require credible evidence that the poor and middle class also benefit from globalization. In this big data age, researchers should heed indicators of real-time measures of support for global free trade. The open question concerns what factors could increase the support for such globalization. The ongoing covid-19 crisis has highlighted the cost of globalization as nations shut down international air travel to protect their citizens from the spread of the virus.

Conclusion: Human Capital Fuels Adaptation

Climate change poses challenges to our quality of life, health, comfort, and productivity. It confronts firms with new challenges for planning logistics, production, and protection of workers and physical assets. Climate change raises the probability of horrible natural disasters taking place.

Leading media outlets such as the *New York Times* and environmental leaders such as U.S. Representative Alexandria Ocasio-Cortez and climate activist Greta Thunberg warn about the tough days ahead posed by climate change. Although it is overly dramatic to say that we have "twelve years left," such messaging focuses our attention on this scary emerging risk.[1] As more people learn about the new climate risks, this may increase the demand for collective action such as enacting carbon taxes. In a democracy, the challenge in enacting such legislation is that those who will suffer an income loss due to this new regulation have an incentive to oppose the regulation. In the case of carbon pricing, fossil fuel consumers, people employed in the fossil fuel industry, and those who own real estate in areas (such as Texas) associated with fossil fuel extraction all lose wealth due to a carbon tax. These individuals are aware of this fact, and they tend to oppose such regulation.[2]

As greenhouse gas emissions rise, we face greater climate change risk. Yet each day we grow stronger with respect to our capacity to cope with this risk. One metric of this success is the fact that the death toll from natural disasters has declined over time. In these chapters we have seen how market forces help us to adapt to climate change risk and thus protect our standard of living. Through rising human capital and experimentation, we are becoming increasingly resilient in the face of the scary threat we now face.

The Bet Round II

The possibility that we are growing ever better at adapting to climate change evokes an older debate. Back in 1980, a prominent bet between two distinguished academics made the national news. This bet crystallized an important debate between an ecologist, Paul Ehrlich, and an economist, Julian Simon. They debated whether the growth of the world's population would harm our standard of living.[3] Ehrlich argued that a growing population would increase demand for scarce resources and would bring about a Malthusian collapse as resources per person would plummet. To appreciate Ehrlich's point, imagine if all 7.4 billion people on the planet grew rich enough to buy a car powered by fossil fuels and each drove that car ten thousand miles each year. If each car's fuel economy is twenty-five miles per gallon, then the aggregate demand for gasoline each year would be 2.96 trillion gallons of gasoline. At this rate of consumption, the world's remaining oil supply would dry up. In the original bet, Ehrlich did not explicitly model behavioral change. Instead, he posited a mechanical model such that as more people earn more money, this simply leads to greater resource consumption. This logic chain resembles the arguments advanced in Jared Diamond's book *Collapse*.

Simon countered by focusing on the role that market price signals play in directing behavioral change. He argued that prices would rise to reflect the rising scarcity of resources, and this would trigger conservation and supply-side innovation to search for substitutes (perhaps the electric car). Simon also argued that population growth raises the possible set of innovators (the next Thomas Edison or Elon Musk) whose ideas would increase the abundance of resources. Ehrlich bet that natural resource prices would rise while Simon bet that prices would fall as human ingenuity would identify substitutes and people would respond to the incentives embodied in price signals. Simon won that bet.

The ongoing climate change adaptation challenge poses a similar high-stakes contest. Given that the world has finite land and many people, what regions will remain livable as climate change grows worse? An extension of Simon's logic posits that our growing human capital will allow us to discover many ways to offset the new environmental pressures we will face. Critics will ask whether we have sufficient time to discover these innovations. Innovators are always looking for the next big profit opportunity. The anticipation of future demand creates an incentive today to research potential solutions. If enough entrepreneurs enter this competition, the probability of a significant breakthrough rises sharply.

In a world featuring well-funded venture capitalists seeking the next unicorn, those young entrepreneurs who foresee the challenges have strong incentives to launch their firm as they seek to be the next Google. To provide just one example, in 2018 the water and waste services group Suez received the Global Desalination Award from the Global Water Awards.[4] This firm's efforts are increasing Egypt's water supply. While I am not qualified to judge the future of Suez, this example highlights how new firms seize emerging challenges because this creates an opportunity for them.

Although we must avoid wishful thinking that technological advance alone will protect us, the technological frontier is shifting out thanks to the rise of the global middle class seeking new products to help them to cope with new risks and financed by the global capital market. Access to the Internet gives billions of people access to real-time information about emerging challenges.

An important challenge for adaptation optimists is to recognize the behavioral economist's critique. This school of thought argues that many of us are not properly processing the new signals arriving about Mother Nature's dynamics. In this polarized news era, our inability to discern true signals of emerging risks leaves us unprepared because of the cacophony of different voices all claiming to be experts. At the same time, there is a rising distrust of experts who appear to have a political slant. Throughout this book, I have discussed the importance of self-awareness. If people are aware that they suffer from the biases that the behavioral economists have noted, then there are many strategies that they can engage in to limit their risk exposure. For example, people who know that they may be overoptimistic about their decisions could first choose to rent in a given area instead of taking out a mortgage. Such a cautious strategy would protect the overoptimists from regretting jumping in by purchasing a piece of real estate that faces significant future risk.

Adaptation can be accelerated or muffled depending on national policy. Through the National Science Foundation, the federal government has an edge both in collecting real-time information about emerging risks and in financing basic research. Such public goods investments help to accelerate adaptation. Other national policies, such as subsidizing insurance and farm activity, have the unintended consequence of slowing adaptation. The interplay between national government policy and the private sector will play a crucial role going forward in determining the pace of adaptation progress.

In 2020, we are adapting to the new risks posed by the coronavirus. We face both contagious disease risk from the virus and fiscal risk associated with the economic shutdown. In the midst of the pandemic, the federal government has changed its rules and policies several times, and this has contributed to uncertainty and planning challenges for private sector firms and households. The private sector can play a more productive role in facilitating adaptation to a shock (whether contagious disease or a natural disaster) if federal policy creates credible incentives for experimentation, innovation, and investment.

The Role of Human Capital in Facilitating Adaptation

Human capital represents our skill in solving problems. Such human capital makes us more productive in every facet of life. Learning begets learning, and skill begets skill. We build up resilience skills by experimenting and trying out strategies proposed by our friends and things we have learned from the Internet. This trial-and-error approach eventually yields new and surprising ways to protect ourselves. All over the world, different people are now confronted with the climate change threat. This creates strong incentives to experiment and to learn from others. Ideas are public goods. Once an innovator figures out a strategy, such blueprints can be replicated elsewhere quite cheaply (think of the low cost of mass-producing generic drugs).

Increased investment in education is taking place, and such investment translates into greater human capital for individuals, nations, and society as a whole. The price of acquiring human capital has fallen as Internet access has spread (think of the Khan Academy and Google's search possibilities). Large nations such as China and India are making enormous investments in expanding education. As billions of people receive increased educational opportunities, this leads to a huge increase in the supply of talent. More educated people are better at identifying new challenges, solving problems, and demanding solutions that can be delivered by both the private sector and our elected officials. More educated people are better able to understand complex problems and to be more patient and thus to engage more on long-term challenges.[5]

Our talent at performing a given task can be built up over time. A great basketball player may be born with talent but must practice for thousands of hours to build up that skill. When a baby is born, a new parent will want to be a good parent but will not know how to be a good parent. Such skill must be acquired through time and effort.

Human capital fuels our understanding of the climate change challenge. Given the importance of the climate science challenge, young people are entering the field to conduct new research. More top universities now have interdisciplinary doctoral programs investigating key climate science issues. Google's ability to organize information guarantees that those who search for information will find useful information. Access to the Internet increases the likelihood that such individuals can access such information at low cost. I think back to when I was a teenager and the inconvenience I faced to take a trip to the local library (located two miles from my house) if there was some issue that I wanted to research. Today, my son simply types at a computer and Google yields the answer much faster than I could thumb through the Dewey Decimal System of the library catalog and then track down the relevant book or newspaper microfiche. Having such easy access to information accelerates learning and discovery.

Throughout this book, I have discussed unknown unknowns related to climate risks. Such unknown unknowns pose profound adaptation threats because we cannot invest resources to prepare for a threat that we do not anticipate. Imagination pares down this set as those people with more imagination are more aware of what is possible. For generations, readers have loved reading science fiction. Authors such as Jules Verne and Isaac Asimov have expanded our imagination with their visions of our future. We live at a time of unprecedented literacy rates and widespread access to information. At a time when there are billions of people who can read, this expands the marketplace for new ideas. This potential market rewards creative authors and reduces the count of unexplored questions, and this helps to transform unknown unknowns into known unknowns. One example is the wide interest in the book *The Uninhabitable Earth*.[6] This scary book is meant to stretch out our imagination so that we can be aware of frightening climate scenarios. This ever increasing awareness of past unknowns distinguishes us from other creatures. After the terrorist attacks of 9/11, many people asked why the US security agencies did not anticipate such an event. Some point to groupthink and the weak incentives in the national security agencies to follow the lead of Steve Jobs to think differently.

In the face of ambiguous risk, decision makers ranging from Jeff Bezos at Amazon to the mayor of Boston will have to decide how prudent they want to be in planning for worst-case scenarios. Asymmetries arise with respect to how costly different mistakes become. Consider the simple example of carrying an umbrella when you do not know if it will rain. Is it more costly for you to carry the umbrella around when it turns out to be sunny? Or is it more costly to have no umbrella when it rains hard? Each decision maker will have to confront this tradeoff in many varied settings. Homeowners will have to decide whether putting their homes on stilts is worth it. People in Berkeley will have to decide whether to invest in air conditioning. These small investment decisions together are the underpinnings of the overall economy's resilience.

Of Mice and Men

All creatures seek to be safe and comfortable. The search for safety and comfort will become more difficult in an increasingly risky world. Unlike us, birds and animals cannot use markets to produce comfort, health, and safety. Instead, they use their time to find food and play defense such as to build a nest. In their simple economies, there are no economies of scale in the production of durable goods such as the mass production of cheap air conditioners. These creatures do not have the imagination to anticipate emerging challenges. Through evolutionary forces, these other creatures have learned to cope within a narrow climate niche. As the climate changes, birds will try to fly to higher, cooler points, but there are limits to how high or far they can fly. Unlike humans, they have fewer adaptation strategies. Our ability to imagine and anticipate ugly future scenarios gives us lead time to prepare.

One way to appreciate the role of human capital is to consider the adaptation challenge that other creatures now face. Consider the example of housing.

> Extreme weather events, such as unusually hot or dry conditions, can cause death by exceeding physiological limits, and so cause loss of population. Survival will depend on whether or not susceptible organisms can find refuges that buffer extreme conditions. Microhabitats offer different microclimates to those found within the wider ecosystem, but do these microhabitats effectively buffer extreme climate events relative to the physiological requirements of the animals that frequent them? We collected temperature data from four common microhabitats (soil, tree holes, epiphytes, and vegetation) located from the ground to canopy in primary rainforests in the Philippines. Ambient temperatures were monitored from outside of each microhabitat and from the upper forest canopy, which represent our macrohabitat controls. We measured the critical thermal maxima (CTmax) of frog and lizard species, which are thermally sensitive and inhabit our microhabitats.[7]

The authors of this study conclude that such homes do shield animals from climate extremes. This example from the natural world is the equivalent of installing heating and air conditioning in your house.

Another example of creatures attempting to adapt are kelp. Here is a snippet from a report on new research on kelp beds: "What [the researchers] found is a relatively rare positive story when it comes to ecological studies in a time of accelerating climate change. The abundance of most modern kelp beds along the Washington coast has remained constant over the last century despite a seawater temperature increase of 0.72 degrees Celsius. The few exceptions are kelp beds closest to Puget Sound, Seattle, and Tacoma."[8]

Modern ecology research examines the adaptation capacity of different creatures.[9] New research finds that penguins species differ with respect to their ability to adapt. The foraging gentoo penguin, for example, has proved to be nimbler in finding food than the chinstrap penguin.[10] Nature is unforgiving when creatures cannot access markets for food and other key survival and quality of life inputs.

Creatures that can move at higher speeds will have a greater menu of spatial locations to look for food. This ecology research demonstrates that other creatures have a more limited set of adaptation strategies than we have. They do not have access to airplanes to fly elsewhere or to borrow money to move to higher ground. They do not have medicines or governments. Human capital fuels our capacity to cope with new challenges. As billions of people all around the world face challenges of extreme heat, drought, and sea level rise, this creates the aggregate demand that provides an incentive for entrepreneurs to devise innovative solutions that reduce these threats. Due to such innovation, we have an ever increasing set of coping strategies. A social justice question arises concerning whether poor people can afford these products. An optimistic answer is that global supply chains connecting nations that are good at mass production, such as China, open up the possibility that the price of such goods will decline due to specialization and competition in output markets.

An open area for future research between scientists and social scientists focuses on how the adaptation of nonhumans affects people. Biology research has studied the rise of the fungus *Candida auris* as a human pathogen and has traced its spatial distribution to rising temperatures that allow it to thrive.[11] It causes infections for vulnerable people and can kill them. This is an example of a living entity that appears to be adapting to emerging climate patterns in ways that pose new risks for people. Understanding these feedback loops focused on how different organisms cope with change will help us to anticipate new public health challenges that will emerge in our hotter world.

Who Will Have Trouble Adapting?

We differ with respect to our ability to cope with change. The poor have fewer affordable coping strategies and thus will face more risk. Climate change will increase inequality. Recognizing this concern, I have discussed several strategies to reduce poverty and help the poor protect themselves through markets and a safety net that does not distort the incentive to work and save. Given that economics is an empirical social science, the question arises whether the poor now face less damage from climate shocks than they used to and whether in the near future they will face even less risk when extreme heat and other climate shocks occur. The evidence of the recent typhoons in India highlight that the death toll from natural disasters in the developing world is declining due to improved access to information technology. Declining prices for proadaptation products helps the poor to become more resilient to risk.

An open scientific question concerns what incentives could be offered to help the poor further adapt to the new challenges we face. The pathway for identifying such incentives relies on experimentation. The 2019 Nobel Prize in Economics was given to three economists who use field experiments to improve the quality of life of people in the developing world. Their core approach embodies modesty. The research team openly acknowledges that they "know that they do not know" what treatments are cost effective for reducing poverty. This team's core research approach is to use randomization to assign people to either the treatment group or the control group. Since the outcome of the coin flip is random, any subsequent gains enjoyed by those who are treated (relative to the gains over time observed for the control group) must be due to the treatment. This "before/after" comparison for the treatment group versus the control group (those who did not receive the treatment) is a powerful tool for learning "what works." This same approach can be used to help the disadvantaged reduce their climate risk exposure.

Tracking Our Standard of Living as the Climate Changes

Economists continue to debate how we measure progress. Put simply, will the average person born in the year 2020 have a much better life than the average person born in 1970 or in 1920? If time machines existed and if we had a market for accessing such time machines, then an economist would look at the market price of a time machine. This would be a clear signal that life is getting better if one had to pay a high price to be born in 2020 versus 1970.

The 1987 Nobel Laureate Robert Solow argued that our economy has achieved sustainable growth if each subsequent generation has a better quality of life than the previous generation.[12] Over the past twenty years, much progress has been made in the war on cancer, but global atmospheric carbon dioxide levels have increased by 20 parts per million. On net, based on Solow's criteria, has sustainability gone up or down? Over the next few decades, will sustainability rise or fall?

Macroeconomic climate research concludes that climate change may cost the world economy trillions in lost income over the next few decades.[13] This enormous cost prediction is based on the fact that extreme heat in the recent past has been associated with lower economic growth and the prediction that climate change will increase extreme heat.

Our individual and collective ability to devise new coping strategies means that historical statistical relations measuring cause and effect overstate the future effects induced by climate change. This core adaptation hypothesis must be continuously evaluated. Evidence that would push me to reject the adaptation hypothesis would include whether we observe over the next few decades that the death toll from natural disasters increases over time; whether the population's health continues to suffer during episodes of extreme heat; and whether firms suffer large productivity losses in the heat and from natural disasters.

The rise of big data offers us a steady stream of new data to judge how people, governments, and firms are adapting over time. The data generated by real-time Internet platforms such as Facebook, Amazon, Twitter, and Google, along with cell phone indicators, offer the possibility of observing in real time how different Americans are coping with new risks. Such individual-level standard of living indicators reveal whether our grandchildren's quality of life continues to improve.

Notes

Introduction

1. ABC7News, "Miley Cyrus, Gerard Butler among Celebrities to Lose Homes."
2. Lopez, "It Wasn't Just the Rich Who Lost Homes."
3. Wolfram, Shelef, and Gertler, "How Will Energy Demand Develop?"; Gertler et al., "Demand for Energy-Using Assets."
4. European Commission, "Paris Agreement."
5. "World Is Losing the War against Climate Change."
6. Metcalf, *Paying for Pollution*.
7. Plumer, "Carbon Dioxide Emissions Hit Record in 2019."
8. Nordhaus, "To Slow or Not to Slow"; Weitzman, "What Is the 'Damages Function' for Global Warming?"
9. Park et al., "Heat and Learning."
10. Hansen, "Nobel Lecture."
11. Kahn, "Climate Change Adaptation Will Offer a Sharp Test."
12. Chetty et al., "Is the United States Still a Land of Opportunity?"
13. Rao, "Climate Change and Housing."
14. Burke and Emerick, "Adaptation to Climate Change."
15. Clemens, "Economics and Emigration."

1 A Microeconomics Perspective on Climate Science Prediction

1. Lakoff, *Unprepared*.
2. Bauer, Thorpe, and Brunet, "Quiet Revolution of Numerical Weather Prediction."
3. Ebert et al., "Progress and Challenges in Forecast Verification"; Stensrud et al., "Progress and Challenges with Warn-on-Forecast."
4. Dessai et al., "Do We Need Better Predictions?"
5. Chinoy, "Places in the US Where Disaster Strikes."

6. Federal Emergency Management Agency, "Disaster Declarations by Year."

7. Wuebbles et al., "Executive Summary."

8. Becker and Lewis, "Interaction between the Quantity and Quality of Children."

9. Weitzman, "Modeling and Interpreting Economics of Catastrophic Climate Change."

10. Curry, "Climate Uncertainty and Risk."

11. Macroeconomic research also seeks to estimate how climate change affects our economy. See Hansen and Brock, "Wrestling with Uncertainty in Climate Economic Models."

12. Hsiang et al., "Estimating Economic Damage from Climate Change"; Hsiang, "Climate Econometrics"; Baylis, "Temperature and Temperament."

13. Heyes and Saberian, "Temperature and Decisions."

14. Zheng et al., "Air Pollution Lowers Chinese Urbanites' Expressed Happiness."

15. Kahn and Li, "Effect of Pollution and Heat."

16. Kahn et al., "Long-Term Macroeconomic Effects of Climate Change."

17. Jagannathan, "High Temperatures Can Lead to More Violent Crime."

18. Burke et al., "Warming Increases Risk of Civil War in Africa."

19. Lucas, "Econometric Policy Evaluation."

20. Thaler, *Misbehaving*.

21. Romm, *Climate Change*.

22. Nobel Prize, Sveriges Riksbank Prize in Economic Sciences in Memory of Alfred Nobel 1978.

23. Meeuwis et al., "Belief Disagreement and Portfolio Choice"; Gennaioli and Shleifer, *Crisis of Beliefs*.

24. Benjamin, Brown, and Shapiro, "Who Is 'Behavioral'?"; Grinblatt, Keloharju, and Linnainmaa, "IQ and Stock Market Participation"; Grinblatt, Keloharju, and Linnainmaa, "IQ, Trading Behavior, and Performance."

25. DellaVigna and Kaplan, "Fox News Effect."

26. Nordhaus, *Climate Casino*.

27. Becker and Murphy, *Social Economics*.

28. Malmendier and Nagel, "Depression Babies."

29. Costa and Kahn, "Changes in the Value of Life."

2 Daily Quality of Life

1. Ritchie and Roser, "Natural Disasters"; Kahn, "Death Toll from Natural Disasters"; Formetta and Feyen, "Empirical Evidence of Declining Global Vulnerability."

2. Costa and Kahn, "Changes in the Value of Life."

3. Gallagher and Hartley, "Household Finance after a Natural Disaster."

4. Bunten and Kahn, "Optimal Real Estate Capital Durability."

5. Molloy, Smith, and Wozniak, "Job Changing and Decline in Long-Distance Migration."

6. Costa and Kahn, "Power Couples."

7. Bütikofer and Peri, "Effects of Cognitive and Noncognitive Skills."

8. Deryugina, Kawano, and Levitt, "Economic Impact of Hurricane Katrina."

9. Egan, "Seattle on the Mediterranean."

10. Rohwedder and Willis, "Mental Retirement"; Smith, McArdle, and Willis, "Financial Decision Making and Cognition."

11. Compton and Pollak, "Proximity and Coresidence of Adult Children and Their Parents."

12. Glaeser, Laibson, and Sacerdote, "Economic Approach to Social Capital."

13. Dingel and Neiman, "How Many Jobs Can Be Done at Home?"

14. Sun, Kahn, and Zheng, "Self-Protection Investment."

15. Simon et al., "Ozone Trends across the United States"; Bento, Mookerjee, and Severnini, "New Approach to Measuring Climate Change."

16. Baylis, "Temperature and Temperament."

17. Zheng et al., "Air Pollution Lowers Chinese Urbanites' Expressed Happiness."

18. Burke et al., "Higher Temperatures Increase Suicide Rates."

19. Constine, "Facebook Rolls Out AI."

20. Bobb et al., "Heat-Related Mortality and Adaptation"; Barreca et al., "Adapting to Climate Change"; Burgess et al., "Weather, Climate Change and Death in India."

21. Petkova, Gasparrini, and Kinney, "Heat and Mortality in New York City."

22. Lo et al., "Increasing Mitigation Ambition."

23. Zheng and Kahn, "New Era of Pollution Progress in Urban China?"

24. Masood, "Dozens Die in Karachi from Relentless Heat."

25. Davis and Gertler, "Contribution of Air Conditioning Adoption."

26. Holland et al., "Environmental Benefits from Driving Electric Vehicles?"

27. Kahn and Zheng, "Blue Skies over Beijing."

28. Heutel, Miller, and Molitor, "Adaptation and Mortality Effects."

29. KB Home, "Energy Efficient Homes."

30. Liu et al., "Particulate Air Pollution from Wildfires"; Liu et al., "Future Respiratory Hospital Admissions."

31. Alen, "Alen BreatheSmart FIT50 True HEPA Air Purifier."

32. Deschênes and Moretti, "Extreme Weather Events."

33. Schwartz, "Canada's Outdoor Rinks Are Melting."

34. Mims, "Scramble for Delivery Robots."

35. US Department of Agriculture, Economic Research Service, "Food Expenditure Series."

36. Yeginsu, "KFC Will Test Vegetarian 'Fried Chicken.'"

37. Acemoglu and Linn, "Market Size in Innovation."

38. Kahn and Zhao, "Impact of Climate Change Skepticism on Adaptation."

39. Koslov, "Avoiding Climate Change."

40. Gates, "Can This Cooler Save Kids from Dying?"

41. Kremer, "Creating Markets for New Vaccines."

42. Boskin, "Toward a More Accurate Measure."

43. Barreca et al., "Adapting to Climate Change."

3 Protecting the Poor

1. Kishore et al., "Mortality in Puerto Rico."
2. Deryugina, "Fiscal Cost of Hurricanes."
3. Hendricks, "Right-Wing Case for Basic Income."
4. Mulligan, "Redistribution Recession."
5. Becker and Mulligan, "Endogenous Determination of Time Preference."
6. Currie, "Healthy, Wealthy, and Wise"; Currie, Neidell, and Schmieder, "Air Pollution and Infant Health."
7. Heckman Equation.
8. Deming, "Growing Importance of Social Skills."
9. Autor, Dorn, and Hanson, "China Syndrome."
10. Neal, "Industry-Specific Human Capital."
11. DiPasquale and Kahn, "Measuring Neighborhood Investments."
12. Boustan et al., "Effect of Natural Disasters."
13. Glaeser and Gyourko, "Urban Decline and Durable Housing."
14. Zillow, "Paradise, California, Home Values."
15. Quigley and Raphael, "Economics of Homelessness."
16. Flavelle and Popovich, "Heat Deaths Jump in Southwest United States."
17. Renthop, "Building Ages and Rents in New York."
18. US Energy Information Administration, "Air Conditioning in Nearly 100 Million U.S. Homes."
19. Home Depot, "8,000 BTU Portable Air Conditioner with Dehumidifier."
20. US Energy Information Administration, "Household Energy Use in Arizona."
21. SMUD, "Low-Income Assistance and Nonprofit Discount."
22. Rawls, *Theory of Justice.*
23. Banzhaf, Ma, and Timmins, "Environmental Justice."
24. Arango, " 'Turn Off the Sunshine.' "
25. City of Los Angeles, "Beat the Heat."
26. Heilmann and Kahn, "Urban Crime and Heat Gradient."
27. Heller et al., "Thinking, Fast and Slow?"
28. Chetty, Hendren, and Katz, "Effects of Exposure to Better Neighborhoods."
29. Bütikofer and Peri, "Effects of Cognitive and Noncognitive Skills."
30. Friedman, *Capitalism and Freedom.*
31. Chyn, "Moved to Opportunity."
32. Mervosh, "Unsafe to Stay, Unable to Go."
33. Deryugina, Kawano, and Levitt, "Economic Impact of Hurricane Katrina."
34. Henderson, Storeygard, and Deichmann, "Has Climate Change Driven Urbanization in Africa?"
35. Bryan, Chowdhury, and Mobarak, "Underinvestment in a Profitable Technology."
36. Kinnan, Wang, and Wang, "Access to Migration for Rural Households."
37. Card, "Impact of Mariel Boatlift."
38. Saiz, "Room in the Kitchen."

39. Field, "Property Rights and Investment."

40. Feler and Henderson, "Exclusionary Policies in Urban Development."

41. Kumar, Gettleman, and Yasir, "How Do You Save a Million People from a Cyclone?"

42. Costa, "Estimating Real Income"; Hamilton, "Using Engel's Law to Estimate CPI Bias."

43. Barreca et al., "Adapting to Climate Change."

4 Upgrading Public Infrastructure

1. Climate Ready Boston, "Climate Change and Sea Level Rise Projections for Boston"; San Diego Foundation, "San Diego's Changing Climate."

2. Mufson, "Boston Harbor Brings Ashore a New Enemy."

3. Bloomberg and Pope, *Climate of Hope.*

4. American Society of Civil Engineers, "2016 Report Card for Humboldt County's Water Infrastructure."

5. Anderson et al., "Dangers of Disaster-Driven Responses"; Wibbenmeyer, Anderson, and Plantinga, "Salience and Government Provision of Public Goods."

6. Healy and Malhotra, "Myopic Voters and Natural Disaster Policy."

7. Kluijver et al., "Advances in the Planning and Conceptual Design of Storm Surge Barriers"; US Army Corps of Engineers, "NY & NJ Harbor & Tributaries Focus Area Feasibility Study."

8. Barnard, "$119 Billion Sea Wall."

9. Sweet, "Why China Can Build High-Speed Rail So Cheaply."

10. Federal Reserve Board, "Federal Reserve Board Releases Results of Supervisory Bank Stress Tests."

11. American Society of Civil Engineers, "2017 Infrastructure Report Card."

12. Hird, "Political Economy of Pork."

13. Flint, "Riches of Resilience."

14. City of Baltimore, "Bond Issue Questions."

15. Cutler and Miller, "Water, Water Everywhere."

16. Costa and Kahn, "Declining Mortality Inequality."

17. Gartner and Zach, "Fighting Fire with Finance."

18. US Environmental Protection Agency, "DC Water's Environmental Impact Bond."

19. Kolden, "What the Dutch Can Teach Us about Wildfires."

20. Paradise Ridge Fire Safe Council.

21. Con Edison, "Our Energy Vision."

22. Kennedy, "Restoring Power after a Natural Disaster."

23. Li et al., "Networked Microgrids"; National Academies of Sciences, Engineering, and Medicine, "Enhancing the Resilience of the Nation's Electricity System"; Moser and Hart, "Adaptation Blindspot."

24. Bruzgul et al., *Rising Seas and Electricity Infrastructure.*

25. Federal Communications Commission, "FCC Seeks Industry Input."

26. New York State Sea Level Rise Task Force, "Report to the Legislature."

27. Tesla, "Gigafactory."

28. Greenstone, Hornbeck, and Moretti, "Identifying Agglomeration Spillovers."

29. US Environmental Protection Agency, Flood Resilience"; Critical Messaging Association, "Technology That Delivers Reliable Communications."

30. McGeehan, "Report Offers 50 Ways to Avoid Chaos."

31. Jerch, Kahn, and Li, "Efficiency of Local Government"; Holmes, "Effect of State Policies on the Location of Manufacturing."

32. Altshuler and Luberoff, "Mega-Projects."

33. Agence France-Presse, "People of Venice Protest."

34. Berkowitz, "What a Chief Resilience Officer Does."

35. Mika et al., "LA Sustainable Water Project."

36. Matchar, "This Concrete Can Absorb a Flood"; Elkhrachy, "Flash Flood Hazard Mapping."

37. St. John and Phillips, "Despite Fire after Fire, Paradise Continued to Boom."

38. Cragg et al., "Carbon Geography"; Dunlap and McCright, "Widening Gap."

39. Madison, "Elizabeth Warren."

40. Fishman, "System That Actually Worked."

5 Will Climate Change Threaten Economic Productivity?

1. Moretti, *New Geography of Jobs.*

2. Glaeser, *Triumph of the City*; Rosenthal and Strange, "Evidence on the Nature and Sources of Agglomeration Economies."

3. "Heated Debate."

4. Hyde, "Schelling's Focal Points."

5. Henderson, Lee, and Lee, "Scale Externalities in Korea."

6. Wang, "Economic Impact of Special Economic Zones."

7. First Street Foundation.

8. Rosen, "Markets and Diversity."

9. Olick, "Amazon's HQ2 in Queens."

10. Adhvaryu, Kala, and Nyshadham, "Management and Shocks to Worker Productivity."

11. Bloom et al., "Does Management Matter?"

12. Jacobson, LaLonde, and Sullivan, "Earnings Losses of Displaced Workers."

13. Bloom et al., "Modern Management."

14. Kunreuther and Useem, *Mastering Catastrophic Risk.*

15. Bloom et al., "Does Management Matter?"

16. Colt, "Here's How Apple Is Making Its New HQ."

17. Levy, "One More Thing."

18. Zivin and Kahn, "Industrial Productivity in a Hotter World."

19. Dingel and Neiman, "How Many Jobs Can Be Done at Home?"

20. Dingel and Neiman, "How Many Jobs Can Be Done at Home?"

21. Bloom et al., "Does Working from Home Work?"
22. Carney, "Resolving the Climate Paradox."
23. Carvalho et al., "Supply Chain Disruptions"; Carvalho, "From Micro to Macro."
24. Kunreuther and Useem, *Mastering Catastrophic Risk.*
25. Donaldson and Storeygard, "View from Above."
26. Walmart, "Enhancing Resilience in the Face of Disasters."
27. Amazon, "FlexiFreeze Ice Vest, Navy."
28. Cahalan, "American Lobster Bonanza Is About to Go Bust."
29. Livingston, "In Alaska, Climate Change."

6 Protecting Urban Real Estate

1. Zillow, "Home Values."
2. Rao, "Climate Change and Housing."
3. Hoffman, Shandas, and Pendleton, "Effects of Historical Housing Policies."
4. Ortega and Taşpınar, "Rising Sea Levels and Sinking Property Values."
5. Beltrán, Maddison, and Elliott, "Impact of Flooding on Property Prices."
6. Kahn, "Climate Change Adaptation Will Offer a Sharp Test."
7. Bernstein, Gustafson, and Lewis, "Disaster on the Horizon."
8. Bakkensen and Barrage, "Flood Risk Belief Heterogeneity."
9. McCright and Dunlap, "Politicization of Climate Change."
10. Ouazad and Kahn, "Mortgage Finance in the Face of Rising Climate Risk."
11. Brannon and Blask, "Government's Hidden Housing Subsidy."
12. DeBonis, "Congress Passes Flood Insurance Extension."
13. Dixon, Tsang, and Fitts, *Impact of Changing Wildfire Risk on California's Residential Insurance Market.*
14. Flavelle and Plumer, "California Bans Insurers from Dropping Policies."
15. Floodscores.
16. Miller, "Startup Plans to Analyze Rising Sea Levels."
17. Rappaport and Sachs, "The United States as a Coastal Nation."
18. Kahn, "Death Toll from Natural Disasters."
19. Kocornik-Mina et al., "Flooded Cities."
20. Butsic, Hanak, and Valletta, "Climate Change and Housing Prices."
21. Commonwealth of Virginia, Executive Order No. 24, 2018.
22. Schwartz and Fausset, "North Carolina, Warned of Rising Seas."
23. Albouy et al., "Climate Amenities;" Kahn, "Urban Growth and Climate Change."
24. Lee, "Singapore's Founding Father"; Oi, "Welfare Implications of Invention."
25. Hsiang, "Climate Econometrics."
26. Chen, "New Buildings in Flood Zones."
27. City of Phoenix, "Phoenix Growth."
28. Barrage, Lint, and Furst, "Housing Investment, Sea Level Rise."
29. Walters, "Plight of Phoenix."
30. Gilbreath, "How I Survived Scorching Hot Phoenix Summers."
31. Josephson, "Cost of Living in Arizona."

32. Glaeser, Kolko, and Saiz, "Consumer City"; Glaeser and Gottlieb, "Urban Resurgence and Consumer City."

33. Costa and Kahn, "Rising Price of Nonmarket Goods."

34. Bakkensen and Barrage, "Flood Risk Belief Heterogeneity."

35. Sorrel, "This Quaint English House."

36. Caughill, "How to Prevent Your House from Flooding"; Adams, "When Waters Rise."

37. Bloomberg and Pope, *Climate of Hope.*

38. Bandell, "Sea-Level Rise Spurs South Florida Developers."

39. Dastrup et al., "Understanding Solar Home Price Premium"; Kahn and Kok, "Capitalization of Green Labels."

40. Kolden, "What the Dutch Can Teach Us about Wildfires."

41. Prescriptive Data, "About Us."

42. Bunten and Kahn, "Optimal Real Estate Capital Durability."

43. Johnson et al., "Benefit-Cost Analysis of Floodplain Land Acquisition."

44. Genesove and Mayer, "Loss Aversion and Seller Behavior."

7 The Market for Big Data Facilitates Adaptation

1. Rappold et al., "Smoke Sense Initiative."

2. Flavelle, "Almost 2 Million Homes Are at Risk."

3. Zheng et al., "Air Pollution Lowers Chinese Urbanites' Expressed Happiness."

4. Shelgikar, Anderson, and Stephens, "Sleep Tracking, Wearable Technology."

5. Sun, Kahn, and Zheng, "Self-Protection Investment."

6. Goldberg, Macis, and Chintagunta, "Leveraging Patients' Social Networks to Overcome Tuberculosis Underdetection."

7. Wolak, "Do Residential Customers Respond to Hourly Prices?"

8. Carter, "First Presidential Report to the American People."

9. Fowlie et al., "Default Effects and Follow-On Behavior"; Gillan, "Dynamic Pricing, Attention, and Automation."

10. House Committee on Science, Space, and Technology, Testimony of Rich Sorkin.

11. Beatty, Shimshack, and Volpe, "Disaster Preparedness and Disaster Response."

12. Courtemanche et al., "Do Walmart Supercenters Improve Food Security?"

13. Amazon, "Sustain Supply Co. Premium Emergency Survival Bag/Kit."

14. Voelker, "Vulnerability to Pandemic Flu."

15. Galbraith, *Affluent Society.*

16. University of Pennsylvania, "Portland Flood Insurance Study."

17. Bishop, "Jeff Bezos Explains Amazon's Artificial Intelligence."

18. USC Viterbi, "Data Science."

19. Moser, "Array of Things."

20. Ghanem and Zhang, "Effortless Perfection."

21. Cohen, "iPhone App Making NBA Smarter."

22. Echenique and Melgar, "Mapping Puerto Rico's Hurricane Migration."

23. Moore et al., "Rapidly Declining Remarkability of Temperature Anomalies."

24. National Science Foundation, "NSF 2026 Idea Machine!"

25. Kang et al., "Using Google Trends for Influenza Surveillance."

26. Choi and Varian, "Predicting the Present with Google Trends."

8 Reimagining the Real Estate Sector

1. Blanchflower and Oswald, "Does High Home-Ownership Impair the Labor Market?"

2. Fisher and Jaffe, "Determinants of International Home Ownership Rates"; Earley, "What Explains the Differences in Homeownership Rates in Europe?"

3. Foote, Gerardi, and Willen, "Negative Equity and Foreclosure."

4. Campbell, Giglio, and Pathak, "Forced Sales and House Prices."

5. Sinai and Souleles, "Owner-Occupied Housing as a Hedge against Rent Risk."

6. Glaeser, Gyourko, and Saks, "Why Is Manhattan So Expensive?"

7. DiPasquale and Glaeser, "Incentives and Social Capital."

8. Olson, *Logic of Collective Action.*

9. Kirwan and Roberts, "Who Really Benefits from Agricultural Subsidies?"

10. Baylis and Boomhower, "Moral Hazard, Wildfires, and the Economic Incidence of Natural Disasters."

11. Lueck and Yoder, "Clearing the Smoke from Wildfire Policy."

12. Federal Emergency Management Agency, "Participation in the National Flood Insurance Program."

13. Hunn, "FEMA on Track to Pay $11 Billion."

14. Cutter and Emrich, "Moral Hazard, Social Catastrophe."

15. Henderson and Mitra, "New Urban Landscape"; Pashigian and Gould, "Internalizing Externalities."

16. Tiebout, "Pure Theory of Local Expenditures."

17. Fischel, "Economics of Zoning Laws."

18. Kahn and Kok, "Big-Box Retailers and Urban Carbon Emissions."

19. Clarke, "Skyscrapers Face the Apocalypse."

20. Bunten and Kahn, "Optimal Real Estate Capital Durability."

21. Brouwen and Kok, "Economics of Energy Labels."

22. Henderson and Mitra, "New Urban Landscape"; Pashigian and Gould, "Internalizing Externalities."

23. Poterba and Sinai, "Tax Expenditures for Owner-Occupied Housing."

9 Reimagining Laws and Regulations to Facilitate Adaptation

1. Hsieh and Moretti, "Housing Constraints and Spatial Misallocation."

2. Glaeser and Kahn, "Greenness of Cities."

3. Fischel, *Homevoter Hypothesis.*

4. Minneapolis 2040, "Access to Housing."

5. Garner, "Neighborhood Spotlight."

6. Frank, Andresen, and Schmid, "Obesity Relationships with Community Design"; Eid et al., "Fat City."

7. Brinklow, "Four-Story Building in Berkeley Built in Four Days."

8. CPUC FireMap.

9. Moretti, "Fires Aren't the Only Threat to the California Dream."

10. Baylis and Boomhower, "Moral Hazard, Wildfires."

11. Baylis and Boomhower, "Moral Hazard, Wildfires."

12. HealthCare.gov, "Glossary: Minimum Essential Coverage (MEC)."

13. State of California, Seismic Safety Commission, "Homeowner's Guide to Earthquake Safety."

14. Harish, "New Law in North Carolina."

15. First Street Foundation.

16. Singapore Government, "Electronic Road Pricing."

17. Nicas, Fuller, and Arango, "Forced Out by Deadly Fires."

18. Kahn, "Environmental Disasters as Risk Regulation Catalysts?"

19. Gagliarducci, Paserman, and Patacchini, "Hurricanes, Climate Change Policies and Electoral Accountability."

20. Miao, "Are We Adapting to Floods?"

21. Wibbenmeyer, Anderson, and Plantinga, "Salience and Government Provision of Public Goods."

10   Innovation in Agricultural Production

1. Evich, " 'I'm Standing Right Here.' "

2. Barnhart, "Local Farmers Forced to Gamble."

3. Penrod, "Drought Forces Hard Choices."

4. Evich, " 'I'm Standing Right Here.' "

5. Burke and Emerick, "Adaptation to Climate Change."

6. Schlenker and Roberts, "Nonlinear Temperature Effects."

7. Asimov, "Climate Change."

8. Pathak et al., "Climate Change Trends and Impacts."

9. Pathak et al., "Climate Change Trends and Impacts."

10. Severen et al., "Forward-Looking Ricardian Approach."

11. Hseih and Moretti, "Housing Constraints and Spatial Misallocation."

12. Heckman, Ichimura, and Todd, "Matching as an Econometric Evaluation Estimator."

13. Lages et al., "Comparison of Land, Water, and Energy Requirements."

14. US Department of Agriculture, National Agricultural Statistics Service, "Farms and Farmland Numbers."

15. Cooke, "40-Foot Shipping Container Farm."

16. Eddy, "Take Caution in Watering Walnuts."

17. Rangel, "From Corgis to Corn."

18. Lusk, Tack, and Hendricks, "Heterogeneous Yield Impacts."

19. Dobert, "Think GMOs Aren't Regulated?"

20. Barrows, Sexton, and Zilberman, "Agricultural Biotechnology."

21. Food and Agriculture Organization of the United Nations, "Water Uses."

22. Hagerty, "Scope for Climate Adaptation."

23. Yan, "Chinese Team Succeeds in Planting Saltwater Rice."

24. Coghlan, "GM Golden Rice Gets Approval."

25. Christensen et al., "Regional Climate Projections."

26. Entine, "Debate about GMO Safety Is Over."

27. Prakash et al., "Risks and Precautions of Genetically Modified Organisms."

28. Roberts and Schlenker, "Is Agricultural Production Becoming More or Less Sensitive to Extreme Heat?"

29. Margaritoff, "Drones in Agriculture."

30. Moore, " 'Sensors Will Profoundly Change Agriculture Decision-Making.' "

31. Velazco, "Arable's Mark Crop Sensors Give Farmers a Data-Driven Edge."

32. African Risk Capacity, "Introduction."

33. Savvides et al., "Chemical Priming of Plants against Multiple Abiotic Stresses."

34. Donaldson and Storeygard, "View from Above."

35. Peng et al., "Benefits of Seasonal Climate Prediction."

36. Carrasco et al., "Global Economic Trade-Offs."

37. Newman and Bunge, "Tariffs May Crown Corn King Again."

38. Alene and Manyong, "Effects of Education on Agricultural Productivity."

39. Ricepedia, "Global Staple."

40. Food and Agriculture Organization of the United Nations, "Farmer Field School Approach."

41. CNFA, "Farmer-to-Farmer: Southern Africa (Closed 2018)."

42. Gloy, "Farm Sector Debt Continues Higher."

43. African Risk Capacity, "How the African Risk Capacity Works."

44. Burwood-Taylor, "Guide to the Investors Funding the Next Agricultural Revolution."

45. Leclerc, "How Sustainable Is the Agritech Venture Ecosystem?"

46. Environmental Working Group, "United States Farm Subsidy Information."

47. US Department of Agriculture, Farm Credit Administration, "Crop Insurance Covers Most Major Crops."

48. US Department of Agriculture, Farm Credit Administration, "Crop Insurance Covers Most Major Crops."

49. OpenSecrets.org, "Center for Responsive Politics."

50. Becker, "Public Policies, Pressure Groups, and Dead Weight Costs"; Annan and Schlenker, "Federal Crop Insurance."

51. Becker, "Theory of Competition among Pressure Groups."

52. Annan and Schlenker, "Federal Crop Insurance."

53. Lemmon et al., "Rapid Improvement of Domestication Traits."

11 Globalization and International Trade Facilitate Adaptation

1. Hornbeck, "Enduring Impact of the American Dust Bowl."
2. Saiz, "Room in the Kitchen"; Saiz, "Immigration and Housing Rents."
3. Ozden, Wagner, and Packard, "Moving for Prosperity."
4. Auerbach, Gokhale, and Kotlikoff, "Generational Accounting."
5. Becker, "Sell the Right to Immigrate."
6. Moraga and Rapoport, "Tradable Immigration Quotas."
7. Jensen and Miller, "Keepin' 'Em down on the Farm."
8. Posner and Weyl, "Sponsor an Immigrant Yourself."
9. *Refugees*.AI.
10. Hartman, "New Life for Refugees."
11. Glaeser, *World of Cities.*
12. Hutchins, "Carriers Say Mega-Ship Sizes Maxing Out."
13. Bates, *Markets and States in Tropical Africa.*
14. International Transport Forum, "Impact of Mega Ships."
15. Trafton, "Comparing Apples and Oranges."
16. World Bank, "World Bank Catastrophe Bond Transaction."

Conclusion

1. Cummings, "World Is Going to End."
2. Cragg et al., "Carbon Geography."
3. Sabin, *The Bet.*
4. Global Water Awards, "2018 Desalination Company of the Year."
5. Becker and Mulligan, "Endogenous Determination of Time Preference."
6. Wallace-Wells, *Uninhabitable Earth.*
7. Scheffers et al., "Microhabitats Reduce Animal's Exposure to Climate Extremes."
8. Wood, "World War I-Era Maps Help Track History of Kelp Forests."
9. Pinsky et al., "Greater Vulnerability to Warming."
10. McMahon et al., "Divergent Trophic Responses of Sympatric Penguin Species."
11. Casadevall, Kontoyiannis, and Robert, "Emergence of *Candida auris.*"
12. Solow, *Sustainability.*
13. Burke, Davis, and Diffenbaugh, "Large Potential Reduction in Economic Damages."

Bibliography

100 Resilient Cities. https://www.100resilientcities.org/.

ABC7News. "Miley Cyrus, Gerard Butler among Celebrities to Lose Homes in Woolsey Fire." November 12, 2018. https://abc7news.com/miley-cyrus-among-celebs-to-lose-homes-in-woolsey-fire/4657842/.

Acemoglu, Daron, and Joshua Linn. "Market Size in Innovation: Theory and Evidence from the Pharmaceutical Industry." *Quarterly Journal of Economics* 119 (2004): 1049–90.

Adams, Dallon. "When Waters Rise, These Flood-Proof Houses Rise Right with Them." *Digital Trends,* September 9, 2017. https://www.digitaltrends.com/home/flood-proof-homes/.

Adhvaryu, Achyuta, Namrata Kala, and Anant Nyshadham. "Management and Shocks to Worker Productivity." NBER Working Paper No. 25865, May 2019.

African Risk Capacity. "Africa RiskView: Introduction." October 31, 2016. https://www.africanriskcapacity.org/2016/10/31/africa-riskview-introduction/.

———. "How the African Risk Capacity Works." https://www.africanriskcapacity.org/about/how-arc-works/.

Agence France-Presse. "People of Venice Protest over Floods and Cruise Ships." *Guardian,* November 24, 2019.

Albouy, David, et al. "Climate Amenities, Climate Change, and American Quality of Life." *Journal of the Association of Environmental and Resource Economists* 3 (2016): 205–46.

Alen. "Alen BreatheSmart FIT50 True HEPA Air Purifier + HEPA-Pure + White." https://www.alencorp.com/products/alen-breathesmart-fit50-hepa-air-purifier.

Alene, Arega D., and V. M. Manyong. "The Effects of Education on Agricultural Productivity under Traditional and Improved Technology in Northern Nigeria: An Endogenous Switching Regression Analysis." *Empirical Economics* 32 (2007): 141–59.

Alsan, Marcella. "The Effect of the Tsetse Fly on African Development." *American Economic Review* 105 (2015): 382–410.

Altshuler, Alan A., and David E. Luberoff. *Mega-Projects: The Changing Politics of Urban Public Investment.* Washington, DC: Brookings Institution Press, 2003.

Amazon. "FlexiFreeze Ice Vest, Navy." https://www.amazon.com/FlexiFreeze-Ice-Vest-Velcro-Closure/dp/B001P30358.

———. "Sustain Supply Co. Premium Emergency Survival Bag/Kit." https://www.amazon.com/Premium-Emergency-Survival-Bag-Kit/dp/B0722L37PL/.

American Society of Civil Engineers. "2016 Report Card for Humboldt County's Water Infrastructure." https://www.infrastructurereportcard.org/wp-content/uploads/2013/02/ASCE-Humboldt-CA-Report-Card-Water-3.24.16.pdf.

———. "2017 Infrastructure Report Card: State Infrastructure Facts." https://www.infrastructurereportcard.org/state-by-state-infrastructure/.

Anderson, Sarah E., et al. "The Dangers of Disaster-Driven Responses to Climate Change." *Nature Climate Change* 8 (2018): 651–53.

Annan, Francis, and Wolfram Schlenker. "Federal Crop Insurance and the Disincentive to Adapt to Extreme Heat." *American Economic Review* 105 (2015): 262–66.

Arango, Tim. " 'Turn Off the Sunshine': Why Shade Is a Mark of Privilege in Los Angeles." *New York Times,* December 1, 2019.

Asimov, Eric. "Great Bubbly from England, Believe It or Not." *New York Times,* December 20, 2018.

———. "How Climate Change Has Altered the Way Cristal Champagne Is Made." *New York Times,* July 19, 2018.

Auerbach, Alan J., Jagadeesh Gokhale, and Laurence J. Kotlikoff. "Generational Accounting: A Meaningful Way to Evaluate Fiscal Policy." *Journal of Economic Perspectives* 8 (1994): 73–94.

Autor, David H., David Dorn, and Gordon H. Hanson. "The China Syndrome: Local Labor Market Effects of Import Competition in the United States." *American Economic Review* 103 (2013): 2121–68.

Bakkensen, Laura, and Lint Barrage. "Flood Risk Belief Heterogeneity and Coastal Home Price Dynamics: Going under Water?" NBER Working Paper No. 23854, 2018.

Bandell, Brian. "Sea-Level Rise Spurs South Florida Developers to Action." *South Florida Business Journal,* July 12, 2018. https://www.bizjournals.com/southflorida/news/2018/07/12/sea-level-rise-spurs-s-fla-developers-to-action.html.

Banzhaf, Spencer, Lala Ma, and Christopher Timmins. "Environmental Justice: The Economics of Race, Place, and Pollution." *Journal of Economic Perspectives* 33 (2019): 185–208.

Barnard, Anne. "The $119 Billion Sea Wall That Could Defend New York . . . or Not." *New York Times,* January 17, 2020.

Barnett, Michael. "A Run on Oil: Climate Policy, Stranded Assets, and Asset Prices." Working Paper, November 2018.

Barnhart, Toria. "Local Farmers Forced to Gamble with Crops as the Poor Weather Persists." *Chillicothe (OH) Gazette,* June 22, 2019.

Barrage, Lint, and Jacob Furst. "Housing Investment, Sea Level Rise, and Climate Change Beliefs." *Economics Letters* 177 (2019): 105–8.

Barreca, Alan, et al. "Adapting to Climate Change: The Remarkable Decline in the U.S. Temperature-Mortality Relationship over the Twentieth Century." *Journal of Political Economy* 124 (2016): 105–59.

Barrett, Scott. "The Incredible Economics of Geoengineering." *Environmental and Resource Economics* 39 (2008): 45–54.

Barrows, Geoffrey, Steven Sexton, and David Zilberman. "Agricultural Biotechnology: The Promise and Prospects of Genetically Modified Crops." *Journal of Economic Perspectives* 28 (2014): 99–120.

Bates, Robert H. *Markets and States in Tropical Africa: The Political Basis of Agricultural Policies.* Berkeley: University of California Press, 2014.

Bauer, Peter, Alan Thorpe, and Gilbert Brunet. "The Quiet Revolution of Numerical Weather Prediction." *Nature* 525 (2015): 47–55.

Baylis, Patrick. "Temperature and Temperament: Evidence from Twitter." *Journal of Public Economics* 184 (2020): 104161.

Baylis, Patrick, and Judson Boomhower. "Moral Hazard, Wildfires, and the Economic Incidence of Natural Disasters." NBER Working Paper No. 26550, December 2019.

Beatty, Timothy K. M., Jay P. Shimshack, and Richard J. Volpe. "Disaster Preparedness and Disaster Response: Evidence from Sales of Emergency Supplies before and after Hurricanes." *Journal of the Association of Environmental and Resource Economists* 6 (2019): 633–68.

Becker, Gary S. "Nobel Lecture: The Economic Way of Looking at Behavior." *Journal of Political Economy* 101 (1993): 385–409.

———. "Public Policies, Pressure Groups, and Dead Weight Costs." *Journal of Public Economics* 28 (1985): 329–47.

———. "Sell the Right to Immigrate—BECKER." Becker-Posner Blog, February 21, 2005. https://www.becker-posner-blog.com/2005/02/sell-the-right-to-immigrate-becker.html.

———. "A Theory of Competition among Pressure Groups for Political Influence." *Quarterly Journal of Economics* 98 (1983): 371–400.

Becker, Gary S., and H. Gregg Lewis. "On the Interaction between the Quantity and Quality of Children." *Journal of Political Economy* 81 (1973): S279–88.

Becker, Gary S., and Casey B. Mulligan. "The Endogenous Determination of Time Preference." *Quarterly Journal of Economics* 112 (1997): 729–58.

Becker, Gary S., and Kevin M. Murphy. *Social Economics: Market Behavior in a Social Environment.* Cambridge, MA: Harvard University Press, 2009.

Beltrán, Allan, David Maddison, and Robert Elliott. "The Impact of Flooding on Property Prices: A Repeat-Sales Approach." *Journal of Environmental Economics and Management* 95 (2019): 62–86.

Benjamin, Daniel J., Sebastian A. Brown, and Jesse M. Shapiro. "Who Is 'Behavioral'? Cognitive Ability and Anomalous Preferences." *Journal of the European Economic Association* 11 (2013): 1231–55.

Bento, Antonio, Mehreen Mookerjee, and Edson Severnini. "A New Approach to Measuring Climate Change Impacts and Adaptation." Working Paper, March 2017.

Berkowitz, Michael. "What a Chief Resilience Officer Does." 100 Resilient Cities, March 18, 2015. http://100resilientcities.org/what-a-chief-resilience-officer-does/.

Bernstein, Asaf, Matthew T. Gustafson, and Ryan Lewis. "Disaster on the Horizon: The Price Effect of Sea Level Rise." *Journal of Financial Economics* 134 (2019): 253–72.

Bishop, Todd. "Jeff Bezos Explains Amazon's Artificial Intelligence Machine Learning Strategy." *Geek Wire,* May 6, 2017. https://www.geekwire.com/2017/jeff-bezos-explains-amazons-artificial-intelligence-machine-learning-strategy/.

Blanchflower, David G., and Andrew J. Oswald. "Does High Home-Ownership Impair the Labor Market?" NBER Working Paper No. 19079, May 2013.

Bleakley, Hoyt. "Disease and Development: Evidence from Hookworm Eradication in the American South." *Quarterly Journal of Economics* 122 (2007): 73–117.

———. "Health, Human Capital, and Development." *Annual Review of Economics* 2 (2010): 283–310.

———. "Malaria Eradication in the Americas: A Retrospective Analysis of Childhood Exposure." *American Economic Journal: Applied Economics* 2 (2010): 1–45.

Bloom, Nicholas, and John Van Reenen. "Why Do Management Practices Differ across Firms and Countries?" *Journal of Economic Perspectives* 24 (2010): 203–24.

Bloom, Nicholas, et al. "Does Management Matter? Evidence from India." *Quarterly Journal of Economics* 128 (2013): 1–51.

———. "Does Working from Home Work? Evidence from a Chinese Experiment." *Quarterly Journal of Economics* 130 (2014): 165–218.

———. "Modern Management: Good for the Environment or Just Hot Air?" *Economic Journal* 120 (2010): 551–72.

Bloomberg, Michael, and Carl Pope. *Climate of Hope: How Cities, Businesses, and Citizens Can Save the Planet.* New York: St. Martin's, 2017.

Bobb, Jennifer F., et al. "Heat-Related Mortality and Adaptation to Heat in the United States." *Environmental Health Perspectives* 122 (2014): 811–16.

Boomhower, Judson. "Drilling Like There's No Tomorrow: Bankruptcy, Insurance, and Environmental Risk." *American Economic Review* 109 (2019): 391–426.

Boskin, Michael J. *Toward a More Accurate Measure of the Cost of Living: Final Report to the Senate Finance Committee from the Advisory Commission to Study the Consumer Price Index.* [Washington, DC]: Advisory Commission to Study the Consumer Price Index, 1996.

Boskin, Michael J., et al. "Consumer Prices, the Consumer Price Index, and the Cost of Living." *Journal of Economic Perspectives* 12 (1998): 3–26.

Boustan, Leah Platt, et al. "The Effect of Natural Disasters on Economic Activity in US Counties: A Century of Data." *Journal of Urban Economics* 118 (2020): 103257.

Brannon, Ike, and Ari Blask. "The Government's Hidden Housing Subsidy for the Rich." *Politico,* August 8, 2017. https://www.politico.com/agenda/story/2017/08/08/hidden-subsidy-rich-flood-insurance-000495.

Brinklow, Adam. "Four-Story Building in Berkeley Built in Four Days." Curbed San Francisco, August 6, 2018. https://sf.curbed.com/2018/8/6/17656118/fast-apartment-residential-building-berkeley-patrick-kennedy-prefab.

Brooks, Nancy, and Rajiv Sethi. "The Distribution of Pollution: Community Characteristics and Exposure to Air Toxics." *Journal of Environmental Economics and Management* 32 (1997): 233–50.

Brounen, Dirk, and Nils Kok. "On the Economics of Energy Labels in the Housing Market." *Journal of Environmental Economics and Management* 62 (2011): 166–79.

Bruzgul, Judsen, et al. *Rising Seas and Electricity Infrastructure: Potential Impacts and Adaptation Actions for San Diego Gas and Electric.* A Report for California's Fourth Climate Change Assessment. California Energy Commission, Publication No. CCCA4-CEC-2018-004, August 2018.

Bryan, Gharad, Shyamal Chowdhury, and Ahmed Mushfiq Mobarak. "Underinvestment in a Profitable Technology: The Case of Seasonal Migration in Bangladesh." *Econometrica* 82 (2014): 1671–748.

Bunten, Devin, and Matthew E. Kahn. "Optimal Real Estate Capital Durability and Localized Climate Change Disaster Risk." *Journal of Housing Economics* 36 (2017): 1–7.

Burgess, Robin, et al. "Weather, Climate Change and Death in India." Working Paper, 2017.

Burke, Marshall, W. Matthew Davis, and Noah S. Diffenbaugh. "Large Potential Reduction in Economic Damages under UN Mitigation Targets." *Nature* 557 (2018): 549–53.

Burke, Marshall, and Kyle Emerick. "Adaptation to Climate Change: Evidence from US Agriculture." *American Economic Journal: Economic Policy* 8 (2016): 106–40.

Burke, Marshall, et al. "Higher Temperatures Increase Suicide Rates in the United States and Mexico." *Nature Climate Change* 8 (2018): 723–29.

———. "Warming Increases the Risk of Civil War in Africa." *PNAS* 106 (2009): 20670–74.

———. "Weather and Violence." *New York Times,* August 30, 2013.

Burwood-Taylor, Louisa. "A Guide to the Investors Funding the Next Agricultural Revolution." *AgFunderNews,* October 6, 2016. https://agfundernews.com/a-guide-to-investors-funding-the-next-agricultural-revolution.html.

Bütikofer, Aline, and Giovanni Peri. "The Effects of Cognitive and Noncognitive Skills on Migration Decisions." NBER Working Paper No. 23877, September 2017, rev. September 2018.

Butsic, Van, Ellen Hanak, and Robert G. Valletta. "Climate Change and Housing Prices: Hedonic Estimates for Ski Resorts in Western North America." *Land Economics* 87 (2011): 75–91.

Cahalan, Susannah. "American Lobster Bonanza Is About to Go Bust." *New York Post,* June 9, 2018.

Campbell, John Y., Stefano Giglio, and Parag Pathak. "Forced Sales and House Prices." *American Economic Review* 101 (2011): 2108–31.

Card, David. "The Impact of the Mariel Boatlift on the Miami Labor Market." *Industrial and Labor Relations Review* 43 (1990): 245–57.

Carney, Mark. "Resolving the Climate Paradox." Bank of England, Arthur Burns Memorial Lecture, Berlin, September 22, 2016.

Carrasco, Luis R., et al. "Global Economic Trade-Offs between Wild Nature and Tropical Agriculture." *PLOS Biology* 15 (2017): e2001657.

Carter, Jimmy. "First Presidential Report to the American People." February 3, 1977. Available at https://www.nytimes.com/1977/02/03/archives/the-text-of-jimmy-carters-first-presidential-report-to-the-american.html.

Carvalho, Vasco M. "From Micro to Macro via Production Networks." *Journal of Economic Perspectives* 28 (2014): 23–48.

Carvalho, Vasco M., et al. "Supply Chain Disruptions: Evidence from the Great East Japan Earthquake." Becker Friedman Institute for Research in Economics Working Paper No. 2017-01, January 2017.

Casadevall, Arturo, Dimitros P. Kontoyiannis, and Vincent Robert. "On the Emergence of *Candida auris:* Climate Change, Azoles, Swamps, and Birds." *mBio* 10 (2019): e01397–19.

Caughill, Daniel. "How to Prevent Your House from Flooding." ValuePenguin, March 24, 2020. https://www.valuepenguin.com/homeowners-insurance/how-to-prevent-home-flooding.

Chen, Stefanos. "New Buildings in Flood Zones." *New York Times,* July 6, 2018.

Chetty, Raj, Nathaniel Hendren, and Lawrence F. Katz. "The Effects of Exposure to Better Neighborhoods on Children: New Evidence from the Moving to Opportunity Experiment." *American Economic Review* 106 (2016): 855–902.

Chetty, Raj, et al. "Is the United States Still a Land of Opportunity? Recent Trends in Intergenerational Mobility." *American Economic Review* 104 (2014): 141–47.

Chinoy, Sahil. "The Places in the U.S. Where Disaster Strikes Again and Again." *New York Times,* May 24, 2018.

Choi, Hyunyoung, and Hal Varian. "Predicting the Present with Google Trends." *Economic Record* 88 (2012): 2–9.

Christensen, J. H., et al. "Regional Climate Projections." In *Climate Change 2007: The Physical Science Basis.* Contribution of Working Group I to the Fourth Assessment Report of the Intergovernmental Panel on Climate Change. Ed. S. Solomon et al. Cambridge: Cambridge University Press, 2007.

Chyn, Eric. "Moved to Opportunity: The Long-Run Effects of Public Housing Demolition on Children." *American Economic Review* 108 (2018): 3028–56.

City of Baltimore. "Bond Issue Questions: GO Bond Loan Authorization." https://planning.baltimorecity.gov/bond-issue-questions.

City of Los Angeles. "Beat the Heat." http://emergency.lacity.org/heat.

City of Phoenix. "Phoenix Growth." https://www.phoenix.gov/budgetsite/Documents/2013Sum%20Community%20Profile%20and%20Trends.pdf.

Clarke, Katherine. "Skyscrapers Face the Apocalypse." *Wall Street Journal*, February 15, 2018.

Clemens, Michael A. "Economics and Emigration: Trillion-Dollar Bills on the Sidewalk?" *Journal of Economic Perspectives* 25 (2011): 83–106.

Climate Ready Boston. "Climate Change and Sea Level Rise Projections for Boston." Boston Research Advisory Group Report, June 1, 2016. https://www.boston.gov/sites/default/files/document-file-12-2016/brag_report_-_final.pdf.

CNFA. "Farmer-to-Farmer: Southern Africa (Closed 2018)." https://www.cnfa.org/program/farmer-to-farmer/.

Coghlan, Andy. "GM Golden Rice Gets Approval from Food Regulators in the US." *New Scientist*, March 30, 2018. https://www.newscientist.com/article/mg23831802-500-gm-golden-rice-gets-approval-from-food-regulators-in-the-us/.

Cohen, Ben. "The iPhone APP Making the NBA Smarter with Artificial Intelligence." *Wall Street Journal*, July 17, 2018.

Colt, Sam. "Here's How Apple Is Making Its New HQ 'The Greenest Building on the Planet.' " *Business Insider*, September 29, 2014. http://www.businessinsider.com/heres-how-apple-is-making-its-new-hq-the-greenest-building-on-the-planet-2014-9.

Commonwealth of Virginia. Executive Order No. 24, 2018: "Increasing Virginia's Resilience to Sea Level Rise and Natural Hazards." https://www.governor.virginia.gov/media/governorvirginiagov/executive-actions/ED-24-Increasing-Virginias-Resilience-To-Sea-Level-Rise-And-Natural-Hazards.pdf.

Compton, Janice, and Robert A. Pollak. "Proximity and Coresidence of Adult Children and Their Parents: Description and Correlates." Michigan Retirement Research Center Research Paper No. 2009-215, October 2009.

Con Edison. "Our Energy Vision: Our Climate Change Resiliency Plan." https://www.coned.com/en/our-energy-future/our-energy-vision/storm-hardening-enhancement-plan.

Constine, Josh. "Facebook Rolls Out AI to Detect Suicidal Posts before They're Reported." *TechCrunch*, November 27, 2017. https://techcrunch.com/2017/11/27/facebook-ai-suicide-prevention/.

Cooke, Lacy. "40-Foot Shipping Container Farm Can Grow 5 Acres of Food with 97% Less Water." InHabitat, July 11, 2017. https://inhabitat.com/40-foot-shipping-container-farm-can-grow-5-acres-of-food-with-97-less-water/.

Costa, Dora L. "Estimating Real Income in the United States from 1888 to 1994: Correcting CPI Bias Using Engel Curves." *Journal of Political Economy* 109 (2001): 1288–310.

Costa, Dora L., and Matthew E. Kahn. "Changes in the Value of Life, 1940–1980." *Journal of Risk and Uncertainty* 29 (2004): 159–80.

———. "Declining Mortality Inequality within Cities during the Health Transition." *American Economic Review* 105 (2015): 564–69.

———. "Power Couples: Changes in the Locational Choice of the College Educated, 1940–1990." *Quarterly Journal of Economics* 115 (2000): 1287–315.

———. "The Rising Price of Nonmarket Goods." *American Economic Review* 93 (2003): 227–32.

Costinot, Arnaud, Dave Donaldson, and Cory Smith. "Evolving Comparative Advantage and the Impact of Climate Change in Agricultural Markets: Evidence from 1.7 Million Fields around the World." *Journal of Political Economy* 124 (2016): 205–48.

Courtemanche, Charles J., et al. "Do Walmart Supercenters Improve Food Security?" *Applied Economic Perspectives and Policy* 41 (2019): 117–98.

CPUC FireMap. https://ia.cpuc.ca.gov/firemap/.

Cragg, Michael I., et al. "Carbon Geography: The Political Economy of Congressional Support for Legislation Intended to Mitigate Greenhouse Gas Production." *Economic Inquiry* 51 (2013): 1640–50.

Critical Messaging Association. "Technology That Delivers Reliable Communications When Disaster Strikes." White Paper. https://www.americanmessaging.net/about/news/ReliableTechnologyWhitePaper.pdf.

Cummings, William. " 'The World Is Going to End in 12 Years If We Don't Address Climate Change,' Ocasio-Cortez Says." *USA Today,* January 22, 2019.

Currie, Janet. "Healthy, Wealthy, and Wise: Socioeconomic Status, Poor Health in Childhood, and Human Capital Development." *Journal of Economic Literature* 47 (2009): 87–122.

Currie, Janet, Matthew Neidell, and Johannes F. Schmieder. "Air Pollution and Infant Health: Lessons from New Jersey." *Journal of Health Economics* 28 (2009): 688–703.

Curry, Judith. "Climate Uncertainty and Risk." *Variations* 16 (2018): 1–7.

Cutler, David M., and Grant Miller. "Water, Water Everywhere: Municipal Finance and Water Supply in American Cities." In *Corruption and Reform: Lessons from America's Economic History,* ed. Edward L. Glaeser and Claudia Golding, 153–84. Chicago: University of Chicago Press, 2006.

Cutter, Susan L., and Christopher T. Emrich. "Moral Hazard, Social Catastrophe: The Changing Face of Vulnerability along the Hurricane Coasts." *Annals of the American Academy of Political and Social Science* 604 (2006): 102–12.

Dastrup, Samuel, et al. "Understanding the Solar Home Price Premium: Electricity Generation and 'Green' Social Status." *European Economic Review* 56 (2012): 961–73.

Davis, Lucas W., and Paul J. Gertler. "Contribution of Air Conditioning Adoption to Future Energy Use under Global Warming." *PNAS* 112 (2015): 5962–67.

DeBonis, Mike. "Congress Passes Flood Insurance Extension, Again Punting on Reforms." *Washington Post,* June 31, 2018.

DellaVigna, Stefano, and Ethan Kaplan. "The Fox News Effect: Media Bias and Voting." *Quarterly Journal of Economics* 122 (2007): 1187–234.

Delmas, Magali A., Matthew E. Kahn, and Stephen L. Locke. "The Private and Social Consequences of Purchasing an Electric Vehicle and Solar Panels: Evidence from California." *Research in Economics* 71 (2017): 225–35.

Deming, David J. "The Growing Importance of Social Skills in the Labor Market." *Quarterly Journal of Economics* 132 (2017): 1593–640.

Deryugina, Tatyana. "The Fiscal Cost of Hurricanes: Disaster Aid versus Social Insurance." *American Economic Journal: Economic Policy* 9 (2017): 168–98.

Deryugina, Tatyana, Laura Kawano, and Steven Levitt. "The Economic Impact of Hurricane Katrina on Its Victims: Evidence from Individual Tax Returns." *American Economic Journal: Applied Economics* 10 (2018): 202–33.

Deschênes, Olivier, and Enrico Moretti. "Extreme Weather Events, Mortality, and Migration." *Review of Economics and Statistics* 91 (2009): 659–81.

Dessai, Suraje, et al. "Do We Need Better Predictions to Adapt to a Changing Climate?" *Eos, Transactions of the American Geophysical Union* 90 (2009): 111–12.

Dingel, Jonathan, and Brent Neiman. "How Many Jobs Can Be Done at Home?" Becker Friedman Institute for Economics, White Paper, June 19, 2020. https://bfi.uchicago.edu/working-paper/how-many-jobs-can-be-done-at-home/.

DiPasquale, Denise, and Edward L. Glaeser. "Incentives and Social Capital: Are Homeowners Better Citizens?" *Journal of Urban Economics* 45 (1999): 354–84.

DiPasquale, Denise, and Matthew E. Kahn. "Measuring Neighborhood Investments: An Examination of Community Choice." *Real Estate Economics* 27 (1999): 389–424.

Dixit, Avinash K., and Robert S. Pindyck. *Investment under Uncertainty.* Princeton, NJ: Princeton University Press, 1994.

Dixon, Lloyd, Flavia Tsang, and Gary Fitts. *The Impact of Changing Wildfire Risk on California's Residential Insurance Market.* California's Fourth Climate Change Assessment. California Natural Resources Agency, Publication No. CCCA4-CNRA-2018-008, 2018.

Dobert, Ray. "Think GMOs Aren't Regulated? Think Again." *Forbes,* December 21, 2015.

Donaldson, Dave, and Adam Storeygard. "The View from Above: Applications of Satellite Data in Economics." *Journal of Economic Perspectives* 30 (2016): 171–98.

Dunlap, Riley E., and Aaron M. McCright. "A Widening Gap: Republican and Democratic Views on Climate Change." *Environment: Science and Policy for Sustainable Development* 50 (2008): 26–35.

Dunlap, Riley E., Aaron M. McCright, and Jerrod H. Yarosh. "The Political Divide on Climate Change: Partisan Polarization Widens in the US." *Environment: Science and Policy for Sustainable Development* 58 (2016): 4–23.

Earley, Fionnuala. "What Explains the Differences in Homeownership Rates in Europe?" *Housing Finance International* 19 (2004): 25–30.

Ebert, E., et al. "Progress and Challenges in Forecast Verification." *Meteorological Applications* 20 (2013): 130–39.

Echenique, Martin, and Luis Melgar. "Mapping Puerto Rico's Hurricane Migration with Mobile Phone Data." CityLab, May 11, 2018. https://www.citylab.com/environment/2018/05/watch-puerto-ricos-hurricane-migration-via-mobile-phone-data/559889/.

Eddy, David. "Take Caution in Watering Walnuts." Growing Produce, July 10, 2012. https://www.growingproduce.com/nuts/take-caution-in-watering-walnuts/.

Egan, Timothy. "Seattle on the Mediterranean." New York Times, July 3, 2015.

Eid, Jean, et al. "Fat City: Questioning the Relationship between Urban Sprawl and Obesity." Journal of Urban Economics 63 (2008): 385–404.

Elkhrachy, Ismail. "Flash Flood Hazard Mapping Using Satellite Images and GIS Tools: A Case Study of Najran City, Kingdom of Saudi Arabia (KSA)." Egyptian Journal of Remote Sensing and Space Science 18 (2015): 261–78.

Entine, Jon. "The Debate about GMO Safety Is Over, Thanks to a New Trillion-Meal Study." Forbes, September 17, 2014.

Evich, Helena B. " 'I'm Standing Right Here in the Middle of Climate Change': How USDA Is Failing Farmers." Politico, October 15, 2019. https://www.politico.com/news/2019/10/15/im-standing-here-in-the-middle-of-climate-change-how-usda-fails-farmers-043615.

Environmental Working Group. "The United States Farm Subsidy Information." https://farm.ewg.org/region.php?fips=00000&progcode=total&yr=2016.

European Commission. "Paris Agreement." https://ec.europa.eu/clima/policies/international/negotiations/paris_en.

Eyer, Jonathan, and Matthew E. Kahn. "Prolonging Coal's Sunset: Local Demand for Local Supply." Regional Science and Urban Economics 81 (2020): 103487.

Federal Communications Commission. "FCC Seeks Industry Input in Review of Wireless Resiliency Framework." November 6, 2018. https://www.fcc.gov/document/fcc-seeks-industry-input-review-wireless-resiliency-framework.

Federal Emergency Management Agency. "Disaster Declarations by Year." https://www.fema.gov/disasters/year.

———. "Participation in the National Flood Insurance Program," March 18, 2019. https://www.fema.gov/participation-national-flood-insurance-program.

Federal Reserve Board. "Federal Reserve Board Releases Results of Supervisory Bank Stress Tests." Press Release, June 21, 2018. https://www.federalreserve.gov/newsevents/pressreleases/bcreg20180621a.htm.

Feler, Leo, and J. Vernon Henderson. "Exclusionary Policies in Urban Development: Under-Servicing Migrant Households in Brazilian Cities." Journal of Urban Economics 69 (2011): 253–72.

Field, Erica. "Property Rights and Investment in Urban Slums." Journal of the European Economic Association 3 (2005): 279–90.

Figlio, David N., and Maurice E. Lucas. "What's in a Grade? School Report Cards and the Housing Market." American Economic Review 94 (2004): 591–604.

First Street Foundation. https://firststreet.org/.

Fischel, William A. The Economics of Zoning Laws: A Property Rights Approach to American Land Use Controls. Baltimore: Johns Hopkins University Press, 1987.

———. The Homevoter Hypothesis: How Home Values Influence Local Government Taxation, School Finance, and Land-Use Policies. Cambridge, MA: Harvard University Press, 2009.

Fisher, Lynn M., and Austin J. Jaffe. "Determinants of International Home Ownership Rates." *Housing Finance International* 18 (2003): 34–42.

Fishman, Charles. "The System That Actually Worked." *Atlantic,* May 6, 2020.

Fisman, Raymond, and Yongxiang Wang. "The Mortality Cost of Political Connections." *Review of Economic Studies* 82 (2015): 1346–82.

Flavelle, Christopher. "Almost 2 Million Homes Are at Risk from Dorian; Most Lack Flood Insurance." *New York Times,* September 6, 2019.

Flavelle, Christopher, and Brad Plumer. "California Bans Insurers from Dropping Policies Made Riskier by Climate Change." *New York Times,* December 5, 2019.

Flavelle, Christopher, and Nadja Popovich. "Heat Deaths Jump in Southwest United States, Puzzling Officials." *New York Times,* August 26, 2019.

Flint, Anthony. "The Riches of Resilience." Lincoln Institute of Land Policy, January 13, 2020. https://www.lincolninst.edu/publications/articles/2020-01-riches-resilience-cities-investing-green-infrastructure-should-developers-foot-bill.

Floodscores. https://floodscores.com/.

Food and Agriculture Organization of the United Nations. "Farmer Field School Approach." www.fao.org/agriculture/ippm/programme/ffs-approach/en/.

———. "Water Uses." *International Rice Commission Newsletter* 48. www.fao.org/nr/water/aquastat/water_use/index.stm.

Foote, Christopher L., Kristopher Gerardi, and Paul S. Willen. "Negative Equity and Foreclosure: Theory and Evidence." *Journal of Urban Economics* 64 (2008): 234–45.

Formetta, Giuseppe, and Luc Feyen. "Empirical Evidence of Declining Global Vulnerability to Climate-Related Hazards." *Global Environmental Change* 57 (2019): 101920.

Fowlie, Meredith, Christopher R. Knittel, and Catherine Wolfram. "Sacred Cars? Cost-Effective Regulation of Stationary and Nonstationary Pollution Sources." *American Economic Journal: Economic Policy* 4 (2012): 98–126.

Fowlie, Meredith, et al. "Default Effects and Follow-On Behavior: Evidence from an Electricity Pricing Program." NBER Working Paper No. 23553, June 2017.

Fox, Karyn M., et al. "Climate Change Adaptation in Ethiopia: Developing a Method to Assess Program Options." In *Resilience: The Science of Adaptation to Climate Change,* ed. Zinta Zommers and Keith Alverson, 253–65. Amsterdam: Elsevier, 2018.

Frank, Lawrence D., Martin A. Andresen, and Thomas L. Schmid. "Obesity Relationships with Community Design, Physical Activity, and Time Spent in Cars." *American Journal of Preventive Medicine* 27 (2004): 87–96.

Frank, Robert H., and Philip J. Cook. *The Winner-Take-All Society: Why the Few at the Top Get So Much More Than the Rest of Us.* New York: Random House, 2010.

Friedman, Milton. *Capitalism and Freedom.* Chicago: University of Chicago Press, 2009.

Gabaix, Xavier. "The Granular Origins of Aggregate Fluctuations." *Econometrica* 79 (2011): 733–72.

Gabaix, Xavier, and David Laibson. "Shrouded Attributes, Consumer Myopia, and Information Suppression in Competitive Markets." *Quarterly Journal of Economics* 121 (2006): 505–40.

Gabaix, Xavier, and Agustin Landier. "Why Has CEO Pay Increased So Much?" *Quarterly Journal of Economics* 123 (2008): 49–100.

Gagliarducci, Stefano, M. Daniele Paserman, and Eleonora Patacchini. "Hurricanes, Climate Change Policies and Electoral Accountability." NBER Working Paper No. 25835, May 2019.

Galbraith, John Kenneth. *The Affluent Society*. 4th ed. New York: Houghton Mifflin Harcourt, 1998.

Gallagher, Justin, and Daniel Hartley. "Household Finance after a Natural Disaster: The Case of Hurricane Katrina." *American Economic Journal: Economic Policy* 9 (2017): 199–228.

Garner, Scott. "Neighborhood Spotlight: Playa Vista Has Master-Planned Convenience, but Developing an Identity Will Take Time." *Los Angeles Times*, June 24, 2017.

Gartner, Todd, and Zach Knight. "Fighting Fire with Finance." *PERC* 37 (2018): https://www.perc.org/2018/07/13/fighting-fire-with-finance/.

Gasparro, Annie, and Jesse Newman. "Six Technologies That Could Shake the Food World." *Wall Street Journal,* October 2, 2018.

Gates, Bill. "Can This Cooler Save Kids from Dying?" GatesNotes, June 13, 2018. https://www.gatesnotes.com/Health/The-big-chill.

Geanakoplos, John. "The Leverage Cycle." *NBER Macroeconomics Annual* 24 (2010): 1–66.

Genesove, David, and Christopher Mayer. "Loss Aversion and Seller Behavior: Evidence from the Housing Market." *Quarterly Journal of Economics* 116 (2001): 1233–60.

Gennaioli, Nicola, and Andrei Shleifer. *A Crisis of Beliefs: Investor Psychology and Financial Fragility*. Princeton, NJ: Princeton University Press, 2018.

Gertler, Paul J., et al. "The Demand for Energy-Using Assets among the World's Rising Middle Classes." *American Economic Review* 106 (2016): 1366–401.

Ghanem, Dalia, and Junjie Zhang. " 'Effortless Perfection': Do Chinese Cities Manipulate Air Pollution Data?" *Journal of Environmental Economics and Management* 68 (2014): 203–25.

Gilbreath, Aaron. "How I Survived Scorching Hot Phoenix Summers." *High Country News,* September 16, 2011. https://www.hcn.org/issues/43.15/how-i-survive-scorching-phoenix-summers.

Gillan, James. "Dynamic Pricing, Attention, and Automation: Evidence from a Field Experiment in Electricity Consumption." Working Paper, November 2017.

Gitig, Diana. "Local Roots: Farm-in-a-Box Coming to a Distribution Center Near You." *Ars Technica,* December 16, 2017. https://arstechnica.com/science/2017/12/local-roots-farm-in-a-box-coming-to-a-distribution-center-near-you/.

Glaeser, Edward. *Triumph of the City: How Urban Spaces Make Us Human*. New York: Pan Macmillan, 2011.

———. "A World of Cities: The Causes and Consequences of Urbanization in Poorer Countries." *Journal of the European Economic Association* 12 (2014): 1154–99.

Glaeser, Edward L., and Joshua D. Gottlieb. "Urban Resurgence and the Consumer City." *Urban Studies* 43 (2006): 1275–99.

Glaeser, Edward L., and Joseph Gyourko. *Rethinking Federal Housing Policy: How to Make Housing Plentiful and Affordable*. Washington, DC: AEI Press, 2008.

———. "Urban Decline and Durable Housing." *Journal of Political Economy* 113 (2005): 345–75.

Glaeser, Edward L., Joseph Gyourko, and Raven Saks. "Why Is Manhattan So Expensive? Regulation and the Rise in Housing Prices." *Journal of Law and Economics* 48 (2005): 331–69.

Glaeser, Edward L., and Matthew E. Kahn. "The Greenness of Cities: Carbon Dioxide Emissions and Urban Development." *Journal of Urban Economics* 67 (2010): 404–18.

Glaeser, Edward L., Jed Kolko, and Albert Saiz. "Consumer City." *Journal of Economic Geography* 1 (2001): 27–50.

Glaeser, Edward L., David Laibson, and Bruce Sacerdote. "An Economic Approach to Social Capital." *Economic Journal* 112 (2002): F437–58.

Global Water Awards. "2018 Desalination Company of the Year." https://globalwaterawards.com/2018-desalination-company-of-the-year/.

Gloy, Brent. "Farm Sector Debt Continues Higher." *Agricultural Economic Insights*, August 7, 2017. https://aei.ag/2017/08/07/farm-sector-debt-continues-higher/.

Gold, Russell. "Harvard Quietly Amasses California Vineyards—and the Water Underneath." *Wall Street Journal*, December 10, 2018.

Goldberg, Jessica, Mario Macis, and Pradeep Chintagunta. "Leveraging Patients' Social Networks to Overcome Tuberculosis Underdetection: A Field Experiment in India." IZA Discussion Paper No. 11942, December 2018.

Goldstein, Allie, et al. "The Private Sector's Climate Change Risk and Adaptation Blind Spots." *Nature Climate Change* 9 (2018): 18–25.

Gorvett, Richard W. "Insurance Securitization: The Development of a New Asset Class." 1999 Casualty Actuarial Society "Securitization of Risk" Discussion Paper Program, 133–73.

Greenstone, Michael, Richard Hornbeck, and Enrico Moretti. "Identifying Agglomeration Spillovers: Evidence from Winners and Losers of Large Plant Openings." *Journal of Political Economy* 118 (2010): 536–98.

Grinblatt, Mark, Matti Keloharju, and Juhani Linnainmaa. "IQ and Stock Market Participation." *Journal of Finance* 66 (2011): 2121–64.

———. "IQ, Trading Behavior, and Performance." *Journal of Financial Economics* 104 (2012): 339–62.

Hagerty, Nick. "The Scope for Climate Adaptation: Evidence from Water Scarcity in Irrigated Agriculture." Job Market Paper, November 21, 2019. http://economics.mit.edu/files/18266.

Hamilton, Bruce W. "Using Engel's Law to Estimate CPI Bias." *American Economic Review* 91 (2001): 619–30.

Hampton, Liz, and Ernest Scheyder. "Houston Still Rebuilding from 2017 Floods as New Hurricane Season Arrives." Reuters, June 1, 2018. http://news.trust.org/item/20180601100017-pzceg/.

Hansen, Lars Peter. "Nobel Lecture: Uncertainty Outside and Inside Economic Models." *Journal of Political Economy* 122 (2014): 945–87.

Hansen, Lars Peter, and William Brock. "Wrestling with Uncertainty in Climate Economic Models." Becker Friedman Institute for Economics Working Paper No. 2019-71, October 2018.

Harish, Alon. "New Law in North Carolina Bans Latest Scientific Predictions of Sea-Level Rise." *ABC News*, August 2, 2012. https://abcnews.go.com/US/north-carolina-bans-latest-science-rising-sea-level/story?id=16913782.

Hartman, Susan. "A New Life for Refugees, and the City They Adopted." *New York Times*, August 10, 2014.

HealthCare.gov. "Glossary: Minimum Essential Coverage (MEC)." https://www.healthcare.gov/glossary/minimum-essential-coverage/.

Healy, Andrew, and Neil Malhotra. "Myopic Voters and Natural Disaster Policy." *American Political Science Review* 103 (2009): 387–406.

"Heated Debate." *Economist*, December 8, 2012.

Heckman, James J., Hidehiko Ichimura, and Petra E. Todd. "Matching as an Econometric Evaluation Estimator: Evidence from Evaluating a Job Training Programme." *Review of Economic Studies* 64 (1997): 605–54.

The Heckman Equation. https://heckmanequation.org.

Heilmann, Kilian, and Matthew E. Kahn. "The Urban Crime and Heat Gradient in High and Low Poverty Areas." NBER Working Paper No. 25961, June 2019.

Heller, Sara B., et al. "Thinking, Fast and Slow? Some Field Experiments to Reduce Crime and Dropout in Chicago." *Quarterly Journal of Economics* 132 (2017): 1–54.

Henderson, Vernon, and Arindam Mitra. "The New Urban Landscape: Developers and Edge Cities." *Regional Science and Urban Economics* 26 (1996): 613–43.

Henderson, Vernon, Todd Lee, and Yung Joon Lee. "Scale Externalities in Korea." *Journal of Urban Economics* 49 (2001): 479–504.

Henderson, Vernon, Adam Storeygard, and Uwe Deichmann. "Has Climate Change Driven Urbanization in Africa?" *Journal of Development Economics* 124 (2017): 60–82.

Hendricks, Scotty. "The Right-Wing Case for Basic Income." *Big Think*, June 11, 2019. https://bigthink.com/politics-current-affairs/negative-income-tax.

Hershfield, Hal E., et al. "Increasing Saving Behavior through Age-Progressed Renderings of the Future Self." *Journal of Marketing Research* 48 (2011): S23–37.

Heutel, Garth, Nolan H. Miller, and David Molitor. "Adaptation and the Mortality Effects of Temperature across US Climate Regions." NBER Working Paper No. 23271, March 2017.

Heyes, Anthony, and Soodeh Saberian. "Temperature and Decisions: Evidence from 207,000 Court Cases." *American Economic Journal: Applied Economics* 11 (2018): 238–65.

Hird, John A. "The Political Economy of Pork: Project Selection at the US Army Corps of Engineers." *American Political Science Review* 85 (1991): 429–56.

Hoffman, Jeremy S., Vivek Shandas, and Nicholas Pendleton. "The Effects of Historical Housing Policies on Resident Exposure to Intra-Urban Heat: A Study of 108 US Urban Areas." *Climate* 8 (2020): 12.

Holian, Matthew J., and Matthew E. Kahn. "Household Demand for Low Carbon Policies: Evidence from California." *Journal of the Association of Environmental and Resource Economists* 2 (2015): 205–34.

Holland, Stephen P., et al. "Are There Environmental Benefits from Driving Electric Vehicles? The Importance of Local Factors." *American Economic Review* 106 (2016): 3700–3729.

Holmes, Thomas J. "The Effect of State Policies on the Location of Manufacturing: Evidence from State Borders." *Journal of Political Economy* 106 (1998): 667–705.

Home Depot. "8,000 BTU Portable Air Conditioner with Dehumidifier." https://www.homedepot.com/p/Arctic-Wind-8-000-BTU-Portable-Air-Conditioner-with-Dehumidifier-AP8018/300043872r.

Hornbeck, Richard. "The Enduring Impact of the American Dust Bowl: Short- and Long-Run Adjustments to Environmental Catastrophe." *American Economic Review* 102 (2012): 1477–507.

Hornbeck, Richard, and Daniel Keniston. "Creative Destruction: Barriers to Urban Growth and the Great Boston Fire of 1872." *American Economic Review* 107 (2017): 1365–98.

Hsiang, Solomon. "Climate Econometrics." *Annual Review of Resource Economics* 8 (2016): 43–75.

Hsiang, Solomon, et al. "Estimating Economic Damage from Climate Change in the United States." *Science* 356 (2017): 1362–69.

Hsieh, Chang-Tai, and Enrico Moretti. "Housing Constraints and Spatial Misallocation." *American Economic Journal: Macroeconomics* 11 (2019): 1–39.

Hummel, Michelle A., Matthew S. Berry, and Mark T. Stacey. "Sea Level Rise Impacts on Wastewater Treatment Systems along the US Coasts." *Earth's Future* 6 (2018): e312.

Hunn, David. "FEMA on Track to Pay $11 Billion in Hurricane Harvey Insurance Claims." *Houston Chronicle*, September 13, 2017.

Hurtado-Díaz, Magali, et al. "Influence of Increasing Temperature on the Scorpion Sting Incidence by Climatic Regions." *International Journal of Climatology* 38 (2018): 2167–73.

Hutchins, Reynolds. "Carriers Say Mega-Ship Sizes Maxing Out, but Doubts Remain." *JOC.com*, May 23, 2017. https://www.joc.com/maritime-news/

container-lines/carriers-say-mega-ship-sizes-maxing-out-doubts-remain_
20170523.html.

Hyde, Tim. "Can Schelling's Focal Points Help Us Understand High-Stakes Negotiations?" *American Economic Association*, January 4, 2017. https://www.aeaweb.org/research/can-schellings-focal-points-help-us-understand-high-stakes-negotiations.

International Transport Forum. "The Impact of Mega Ships." 2015. https://www.itf-oecd.org/sites/default/files/docs/15cspa_mega-ships.pdf.

Ito, Koichiro. "Do Consumers Respond to Marginal or Average Price? Evidence from Nonlinear Electricity Pricing." *American Economic Review* 104 (2014): 537–63.

Jacobson, Louis S., Robert J. LaLonde, and Daniel G. Sullivan. "Earnings Losses of Displaced Workers." *American Economic Review* 83 (1993): 685–709.

Jagannathan, Meera. "High Temperatures Can Lead to More Violent Crime," Study Finds." *New York Post*, June 18, 2019. https://nypost.com/2019/06/18/high-temperatures-can-lead-to-more-violent-crime-study-finds/.

Jensen, Robert, and Nolan H. Miller. "Keepin' 'Em Down on the Farm: Migration and Strategic Investment in Children's Schooling." NBER Working Paper No. 23122, February 2017.

Jerch, Rhiannon, Matthew E. Kahn, and Shanjun Li. "The Efficiency of Local Government: The Role of Privatization and Public Sector Unions." *Journal of Public Economics* 154 (2017): 95–121.

Johnson, Kris A., et al. "A Benefit-Cost Analysis of Floodplain Land Acquisition for US Flood Damage Reduction." *Nature Sustainability* (December 2019): 1–7.

Josephson, Amelia. "The Cost of Living in Arizona." *Smartasset*, May 28, 2019. https://smartasset.com/mortgage/the-cost-of-living-in-arizona.

Kahn, Matthew E. "The Beneficiaries of Clean Air Act Regulation." *Regulation Magazine* 24 (2001): 34.

———. "Climate Change Adaptation Will Offer a Sharp Test of the Claims of Behavioral Economics." *Economists' Voice* 12 (2015): 25–30.

———. "The Death Toll from Natural Disasters: The Role of Income, Geography, and Institutions." *Review of Economics and Statistics* 87 (2005): 271–84.

———. "Environmental Disasters as Risk Regulation Catalysts? The Role of Bhopal, Chernobyl, Exxon Valdez, Love Canal, and Three Mile Island in Shaping US Environmental Law." *Journal of Risk and Uncertainty* 35 (2007): 17–43.

———. "Urban Growth and Climate Change." *Annual Review of Resource Economics* 1 (2009): 333–50.

Kahn, Matthew E., and Nils Kok. "Big-Box Retailers and Urban Carbon Emissions: The Case of Wal-Mart." NBER Working Paper No. 19912, February 2014.

———. "The Capitalization of Green Labels in the California Housing Market." *Regional Science and Urban Economics* 47 (2014): 25–34.

Kahn, Matthew E., and Pei Li. "The Effect of Pollution and Heat on High Skill Public Sector Worker Productivity in China." NBER Working Paper No. 25594, February 2019.

Kahn, Matthew E., and Frank A. Wolak. "Using Information to Improve the Effectiveness of Nonlinear Pricing: Evidence from a Field Experiment." California Air Resources Board, Research Division, March 2013.

Kahn, Matthew E., and Daxuan Zhao. "The Impact of Climate Change Skepticism on Adaptation in a Market Economy." *Research in Economics* 72 (2018): 251–62.

Kahn, Matthew E., and Siqi Zheng. *Blue Skies over Beijing: Economic Growth and the Environment in China.* Princeton, NJ: Princeton University Press, 2016.

Kahn, Matthew E., et al. "Long-Term Macroeconomic Effects of Climate Change: A Cross-Country Analysis." IMF Working Papers, October 2019.

Kang, Min, et al. "Using Google Trends for Influenza Surveillance in South China." *PloS One* 8 (2013): e55205.

KB Home. "Energy Efficient Homes." https://www.kbhome.com/energy-efficient-homes.

Keith, David W. "Geoengineering." *Nature* 409 (2001): 420.

Kennedy, Chad. "Restoring Power after a Natural Disaster: How to Plan for the Worst." *Plant Services,* July 11, 2017. https://www.plantservices.com/articles/2017/es-restoring-power-after-a-natural-disaster/.

Kinnan, Cynthia, Shing-Yi Wang, and Yongxiang Wang. 2018. "Access to Migration for Rural Households." *American Economic Journal: Applied Economics* 10 (2018): 79–119.

Kirwan, Barrett E., and Michael J. Roberts. "Who *Really* Benefits from Agricultural Subsidies? Evidence from Field-Level Data." *American Journal of Agricultural Economics* 98 (2016): 1095–113.

Kishore, Nishant, et al. "Mortality in Puerto Rico after Hurricane Maria." *New England Journal of Medicine* 379 (2018): 162–70.

Kluijver, Maarten, et al. "Advances in the Planning and Conceptual Design of Storm Surge Barriers—Application to the New York Metropolitan Area." *Coastal Structures* (2019): 326–36.

Kocornik-Mina, Adriana, et al. "Flooded Cities." *American Economic Journal: Applied Economics* 12 (2020): 35–66.

Kolden, Crystal. "What the Dutch Can Teach Us about Wildfires." *New York Times,* November 16, 2018.

Koslov, Liz. "Avoiding Climate Change: 'Agnostic Adaptation' and the Politics of Public Silence." *Annals of the American Association of Geographers* 109 (2019): 568–80.

Kremer, Michael. "Creating Markets for New Vaccines. Part I: Rationale." *Innovation Policy and the Economy* 1 (2000): 35–72.

Kumar, Hari, Jeffrey Gettleman, and Sameer Yasir. "How Do You Save a Million People from a Cyclone? Ask a Poor State in India." *New York Times,* May 3, 2019.

Kunreuther, Howard, and Michael Useem. *Mastering Catastrophic Risk: How Companies Are Coping with Disruption.* Oxford: Oxford University Press, 2018.

Lages Barbosa, Guilherme, et al. "Comparison of Land, Water, and Energy Requirements of Lettuce Grown Using Hydroponic vs. Conventional Agricultural Methods." *International Journal of Environmental Research and Public Health* 12 (2015): 6879–91.

Lakoff, Andrew. *Unprepared: Global Health in a Time of Emergency.* Berkeley: University of California Press, 2017.

Larcinese, Valentino, Leonzio Rizzo, and Cecilia Testa. "Allocating the US Federal Budget to the States: The Impact of the President." *Journal of Politics* 68 (2006): 447–56.

Laurent, Jose Guillermo Cedeño, et al. "Reduced Cognitive Function during a Heat Wave among Residents of Non-Air-Conditioned Buildings: An Observational Study of Young Adults in the Summer of 2016." *PLOS Medicine* 15 (2018): e1002605.

Leclerc, Rob. "How Sustainable Is the Agritech Venture Ecosystem?" *Forbes,* August 17, 2016.

Lee, Katy. "Singapore's Founding Father Thought That Air Conditioning Was the Secret to His Country's Success." *Vox,* March 23, 2015. https://www.vox.com/2015/3/23/8278085/singapore-lee-kuan-yew-air-conditioning.

Lemmon, Zachary H., et al. "Rapid Improvement of Domestication Traits in an Orphan Crop by Genome Editing." *Nature Plants* 4 (2018): 766–70.

Lenton, Timothy M. "Early Warning of Climate Tipping Points." *Nature Climate Change* 1 (2011): 201–9.

Levy, Steben. "One More Thing." *Wired Magazine,* May 16, 2017.

Li, Zhiyi, et al. "Networked Microgrids for Enhancing the Power System Resilience." *Proceedings of the IEEE* 105 (2017): 1289–310.

Lippman, Zachary B. "Rapid Improvement of Domestication Traits in an Orphan Crop by Genome Editing." *Nature Plants* 4 (2018): 766–70.

Liu, Jia Coco, et al. "Future Respiratory Hospital Admissions from Wildfire Smoke under Climate Change in the Western US." *Environmental Research Letters* 11 (2016): 124018.

———. "Particulate Air Pollution from Wildfires in the Western US under Climate Change." *Climatic Change* 138 (2016): 655–66.

Livingston, Ian. "In Alaska, Climate Change Is Showing Increasing Signs of Disrupting Everyday Life." *Washington Post,* May 8, 2019.

Lo, Y. T. Eunice, et al. "Increasing Mitigation Ambition to Meet the Paris Agreement's Temperature Goal Avoids Substantial Heat-Related Mortality in US Cities." *Science Advances* 5 (2019): eaau4373.

Lopez, Steve. "It Wasn't Just the Rich Who Lost Homes in the Malibu Area; Is Fire California's Great Equalizer?" *Los Angeles Times,* November 14, 2018.

Lu, Xin, et al. "Unveiling Hidden Migration and Mobility Patterns in Climate Stressed Regions: A Longitudinal Study of Six Million Anonymous Mobile Phone Users in Bangladesh." *Global Environmental Change* 38 (2016): 1–7.

Lucas, Robert E., Jr. "Econometric Policy Evaluation: A Critique." In *Carnegie-Rochester Conference Series on Public Policy*, 1:19–46. Amsterdam: North Holland, 1976.

Ludwig, Jens, and Steven Raphael. "The Mobility Bank: Increasing Residential Mobility to Boost Economic Mobility." Hamilton Project, Brookings Institution, October 2010.

Lueck, Dean, and Jonathan Yoder. "Clearing the Smoke from Wildfire Policy: An Economic Perspective." Property and Environment Research Center, July 21, 2016. https://www.perc.org/2016/07/21/clearing-the-smoke-from-wildfire-policy-an-economic-perspective/.

Lusk, Jayson L., Jesse Tack, and Nathan P. Hendricks. "Heterogeneous Yield Impacts from Adoption of Genetically Engineered Corn and the Importance of Controlling for Weather." NBER Working Paper No. 23519, June 2017.

Maciag, Mike. "Risky Waters." *Governing*, August 3, 2018. http://www.governing.com/topics/transportation-infrastructure/gov-flood-zone-floodplain-development-homes-zoning.html.

Madison, Lucy. "Elizabeth Warren: 'There Is Nobody in This Country Who Got Rich on His Own.' " *CBS News*, September 22, 2011. https://www.cbsnews.com/news/elizabeth-warren-there-is-nobody-in-this-country-who-got-rich-on-his-own/.

Malmendier, Ulrike, and Stefan Nagel. "Depression Babies: Do Macroeconomic Experiences Affect Risk Taking?" *Quarterly Journal of Economics* 126 (2011): 373–416.

Malmendier, Ulrike, and Geoffrey Tate. "CEO Overconfidence and Corporate Investment." *Journal of Finance* 60 (2005): 2661–700.

Margaritoff, Marco. "Drones in Agriculture: How UAVs Make Farming More Efficient." *The Drive*, February 13, 2018. https://www.thedrive.com/tech/18456/drones-in-agriculture-how-uavs-make-farming-more-efficient.

Masood, Salmon. "Dozens Die in Karachi from Relentless Heat." *New York Times*. May 21, 2018.

Matchar, Emily. "This Concrete Can Absorb a Flood." *Smithsonian Magazine*, October 5, 2015.

McClure, Crystal D., and Daniel A. Jaffe. "US Particulate Matter Air Quality Improves Except in Wildfire-Prone Areas." *PNAS* 115 (2018): 7901–6.

McCright, Aaron M., and Riley E. Dunlap. "The Politicization of Climate Change and Polarization in the American Public's Views of Global Warming, 2001–2010." *Sociological Quarterly* 52 (2011): 155–94.

McGeehan, Patrick. "Report Offers 50 Ways to Avoid Chaos That Crippled Kennedy Airport." *New York Times*, May 31, 2018.

McMahon, Kelton W., et al. "Divergent Trophic Responses of Sympatric Penguin Species to Historic Anthropogenic Exploitation and Recent Climate Change." *PNAS* 116 (2019): 25721–27.

Meeuwis, Maarten, et al. "Belief Disagreement and Portfolio Choice." NBER Working Paper No. 25108, September 2018, rev. September 2019.

Mervosh, Sarah. "Unsafe to Stay, Unable to Go: Half a Million Face Flooding Risk in Government Homes." *New York Times,* April 11, 2019.

Metcalf, Gilbert E. "An Equitable Tax Reform to Address Global Climate Change." Discussion Paper 2007-12. Hamilton Project, Brookings Institution, 2007.

———. *Paying for Pollution: Why America Needs a Carbon Tax.* New York: Oxford University Press; 2019.

Miao, Qing. "Are We Adapting to Floods? Evidence from Global Flooding Fatalities." *Risk Analysis* (2018). doi: 10.1111/risa.13245.

Mika, Katie, et al. "LA Sustainable Water Project: Los Angeles City-Wide Overview." Working Paper, UCLA, February 2018.

Miller, Kimberly. "Startup Plans to Analyze Rising Sea Levels to Aid Florida Homeowners." *Palm Beach Post,* July 10, 2018.

Mims, Christopher. "The Scramble for Delivery Robots Is On and Startups Can Barely Keep Up." *Wall Street Journal,* April 25, 2020.

Minneapolis 2040. "Access to Housing." https://minneapolis2040.com/policies/access-to-housing/.

Molloy, Raven, Christopher L. Smith, and Abigail Wozniak. "Job Changing and the Decline in Long-Distance Migration in the United States." *Demography* 54 (2017): 631–53.

Moore, Frances C., et al. "Rapidly Declining Remarkability of Temperature Anomalies May Obscure Public Perception of Climate Change." *PNAS* 116 (2019): 4905–10.

Moore, Mark. " 'Sensors Will Profoundly Change Agriculture Decision-Making.' " *Successful Farming,* December 1, 2017. https://www.agriculture.com/technology/data/sensors-will-profoundly-change-agriculture-decision-making.

Moraga, Jesús Fernández-Huertas, and Hillel Rapoport. "Tradable Immigration Quotas." *Journal of Public Economics* 115 (2014): 94–108.

Morello-Frosch, Rachel, Manuel Pastor, and James Sadd. "Environmental Justice and Southern California's 'Riskscape': The Distribution of Air Toxics Exposures and Health Risks among Diverse Communities." *Urban Affairs Review* 36 (2001): 551–78.

Moreno-Cruz, Juan B., and David W. Keith. "Climate Policy under Uncertainty: A Case for Solar Geoengineering." *Climatic Change* 121 (2013): 431–44.

Moretti, Enrico. "Fires Aren't the Only Threat to the California Dream." *New York Times,* November 3, 2017.

———. *The New Geography of Jobs.* New York: Houghton Mifflin Harcourt, 2012.

Moser, Susanne C., and Juliette Finzi Hart. "The Adaptation Blindspot: Teleconnected and Cascading Impacts of Climate Change in the Electrical Grid and Lifelines of Los Angeles." California Energy Commission, Publication No. CCCA4-CEC-2018-008. 2018.

Moser, Whet. "The Array of Things Is Coming to Chicago." *Chicago Magazine,* September 2, 2016.

Mufson, Steven. "Boston Harbor Brings Ashore a New Enemy: Rising Seas." *Washington Post,* February 19, 2020.

Mulligan, Casey B. *The Redistribution Recession: How Labor Market Distortions Contracted the Economy.* Oxford: Oxford University Press, 2012.

National Academies of Sciences, Engineering, and Medicine. *Enhancing the Resilience of the Nation's Electricity System.* Washington, DC: National Academies Press, 2017.

National Science Foundation. "The NSF 2026 Idea Machine!" https://www.nsf.gov/news/special_reports/nsf2026ideamachine/index.jsp.

Neal, Derek. "Industry-Specific Human Capital: Evidence from Displaced Workers." *Journal of Labor Economics* 13 (1995): 653–77.

Newman, Jesse, and Jacob Bunge. "Tariffs May Crown Corn King Again." *Wall Street Journal,* October 28, 2018.

New York State Sea Level Rise Task Force. "Report to the Legislature." December 2010. https://www.dec.ny.gov/docs/administration_pdf/slrtffinalrep.pdf.

Ngu, Ash, and Sahil Chinoy. "To Help Prevent the Next Big Wildfire, Let the Forest Burn." *New York Times,* November 29, 2018.

Nicas, Jack, Thomas Fuller, and Tim Arango. "Forced Out by Deadly Fires, Then Trapped in Traffic." *New York Times,* November 11, 2018.

Nobel Prize. The Sveriges Riksbank Prize in Economic Sciences in Memory of Alfred Nobel 1978: "Studies of Decision Making Lead to Prize in Economics." Press Release, October 16, 1978. https://www.nobelprize.org/nobel_prizes/economic-sciences/laureates/1978/press.html.

Nordhaus, William D. *The Climate Casino: Risk, Uncertainty, and Economics for a Warming World.* New Haven: Yale University Press, 2013.

———. "To Slow or Not to Slow: The Economics of the Greenhouse Effect." *Economic Journal* 101 (1991): 920–37.

Obradovich, Nick, et al. "Empirical Evidence of Mental Health Risks Posed by Climate Change." *PNAS* 115 (2018): 10953–58.

Oi, Walter Y. "The Welfare Implications of Invention." In *The Economics of New Goods,* ed. Timothy F. Bresnahan and Robert J. Goode, 109–42. Chicago: University of Chicago Press, 1996.

Olick, Diana. "Amazon's HQ2 in Queens Will Be 'Square in the Danger Zone for Frequent Flooding.' " *CNBC,* November 16, 2018. https://www.cnbc.com/2018/11/16/amazon-hq2-in-queens-will-be-in-danger-zone-for-frequent-flooding.html.

Olson, Mancur. *The Logic of Collective Action: Public Goods and the Theory of Groups.* Cambridge, MA: Harvard University Press, 2009.

O'Neil, Cathy. *Weapons of Math Destruction: How Big Data Increases Inequality and Threatens Democracy.* New York: Broadway Books, 2016.

OpenSecrets.org. Center for Responsive Politics. "Lobbying Spending Database." https://www.opensecrets.org/federal-lobbying.

Ortega, Francesc, and Süleyman Taşpınar. "Rising Sea Levels and Sinking Property Values: Hurricane Sandy and New York's Housing Market." *Journal of Urban Economics* 106 (2018): 81–100.

Ouazad, Amine, and Matthew E. Kahn. "Mortgage Finance in the Face of Rising Climate Risk." NBER Working Paper No. 26322, September 2019.

Özden, Çağlar, Matthis Wagner, and Michael Packard. "Moving for Prosperity: Global Migration and Labor Markets." Policy Research Report Overview, World Bank Group. June 2018.

Pacala, S., and R. Socolow. "Stabilization Wedges: Solving the Climate Problem for the Next 50 Years with Current Technologies." *Science* 305 (2004): 968–72.

Paradise Ridge Fire Safe Council. www.paradisefiresafe.org.

Park, R. Jisung, et al. "Heat and Learning." *American Economic Journal: Economic Policy* 12 (2020): 306–39.

Pashigian, B. Peter, and Eric D. Gould. "Internalizing Externalities: The Pricing of Space in Shopping Malls." *Journal of Law and Economics* 41 (1998): 115–42.

Pathak, Tapan B., et al. "Climate Change Trends and Impacts on California Agriculture: A Detailed Review." *Agronomy* 8 (2018): 25.

Peng, Bin, et al. "Benefits of Seasonal Climate Prediction and Satellite Data for Forecasting U.S. Maize Yield." *Geophysical Research Letters* 45 (2018): 9662–71.

Penrod, Emma. "Drought Forces Hard Choices for Farmers and Ranchers in the Southwest." News Deeply: Water Deeply, August 13, 2018. www.newsdeeply. com/water/articles/2018/08/13/drought-forces-hard-choices-for-farmers-and-ranchers-in-the-southwest.

Petkova, Elizaveta P., Antonio Gasparrini, and Patrick L. Kinney. "Heat and Mortality in New York City since the Beginning of the 20th Century." *Epidemiology* 25 (2014): 554–60.

Phelps, Edmund S. *Rewarding Work: How to Restore Participation and Self-Support to Free Enterprise.* Cambridge, MA: Harvard University Press, 2007.

Piketty, Thomas, and Emmanuel Saez. "Income Inequality in the United States, 1913–1998." *Quarterly Journal of Economics* 118 (2003): 1–41.

Pinsky, Malin L., et al. "Greater Vulnerability to Warming of Marine versus Terrestrial Ectotherms." *Nature* 569 (2019): 108–11.

Plumer, Brad. "Carbon Dioxide Emissions Hit a Record in 2019, Even as Coal Fades." *New York Times,* December 3, 2019.

Posner, Eric, and Glen Weyl. "Sponsor an Immigrant Yourself." Politico, February 13, 2018. www.politico.com/magazine/story/2018/02/13/immigration-visas-economics-216968.

Poterba, James, and Todd Sinai. "Tax Expenditures for Owner-Occupied Housing: Deductions for Property Taxes and Mortgage Interest and the Exclusion of Imputed Rental Income." *American Economic Review* 98 (2008): 84–89.

Prakash, Dhan, et al. "Risks and Precautions of Genetically Modified Organisms." *ISRN Ecology* (2011). doi: 10.5402/2011/369573.

Prescriptive Data. "About Us." https://www.prescriptivedata.io/about/.

Quigley, John M., and Steven Raphael. "The Economics of Homelessness: The Evidence from North America." *European Journal of Housing Policy* 1 (2001): 323–36.

Rabin, Matthew. "Inference by Believers in the Law of Small Numbers." *Quarterly Journal of Economics* 117 (2002): 775–816.

Rangel, Gabriel. "From Corgis to Corn: A Brief Look at the Long History of GMO Technology." Science in the News, Harvard University, August 9, 2015. sitn.hms. harvard.edu/flash/2015/from-corgis-to-corn-a-brief-look-at-the-long-history-of-gmo-technology/.

Rao, Khrishna. "Climate Change and Housing: Will a Rising Tide Sink All Homes?" Zillow, June 2, 2017. http://www.zillow.com/research/climate-changeunderwa ter-homes-12890.

Rappaport, Jordan, and Jeffrey D. Sachs. "The United States as a Coastal Nation." *Journal of Economic Growth* 8 (2003): 5–46.

Rappold, A. G., et al. "Smoke Sense Initiative Leverages Citizen Science to Address the Growing Wildfire-Related Public Health Problem." *GeoHealth* 3 (2019): 443–57.

Rathi, Akshat. "Why the New Nobel Laureate Is Optimistic about Beating Climate Change." *Quartz,* October 8, 2018. https://qz.com/1417222/why-new-nobel-laureate-paul-romer-is-optimistic-about-beating-climate-change/.

Rawls, John. *A Theory of Justice.* Cambridge, MA: Harvard University Press, 2009.

*Refugees.*AI. https://www.refugees.ai/.

Renthop. "Building Ages and Rents in New York." August 24, 2017. https://www. renthop.com/studies/nyc/building-age-and-rents-in-new-york.

Reiss, Peter C., and Matthew W. White. "What Changes Energy Consumption? Prices and Public Pressures." *RAND Journal of Economics* 39 (2008): 636–63.

Restuccia, Diego, and Richard Rogerson. "The Causes and Costs of Misallocation." *Journal of Economic Perspectives* 31 (2017): 151–74.

Ricepedia. "The Global Staple." ricepedia.org/rice-as-food/the-global-staple-rice-consumers.

Ringleb, Al H., and Steven N. Wiggins. "Liability and Large-Scale, Long-Term Hazards." *Journal of Political Economy* 98 (1990): 574–95.

Ritchie, Hannah, and Max Roser. "Natural Disasters." *Our World in Data,* 2014, rev. November 2019. https://ourworldindata.org/natural-disasters.

Roberts, Michael J., and Wolfram Schlenker. "Is Agricultural Production Becoming More or Less Sensitive to Extreme Heat? Evidence from US Corn and Soybean Yields." NBER Working Paper No. 16308, August 2010.

Rohwedder, Susann, and Robert J. Willis. "Mental Retirement." *Journal of Economic Perspectives* 24 (2010): 119–38.

Romer, Paul. "Technologies, Rules, and Progress: The Case for Charter Cities." Center for Global Development Essay, March 2010. https://www.cgdev.org/sites/default/files/1423916_file_TechnologyRulesProgress_FINAL.pdf.

Romm, Joseph. *Climate Change: What Everyone Needs to Know.* New York: Oxford University Press, 2018.

Rosen, Sherwin. "Austrian and Neoclassical Economics: Any Gains from Trade?" *Journal of Economic Perspectives* 11 (1997): 139–52.

———. "Markets and Diversity." *American Economic Review* 92 (2002): 1–5.

Rosenthal, Stuart S., and William C. Strange. "Evidence on the Nature and Sources of Agglomeration Economies." In *Handbook of Regional and Urban Economics*, ed. J. Vernon Henderson and Jacques-François Thisse, 4:2119–71. Amsterdam: Elsevier, 2004.

Rossi-Hansberg, Esteban, Pierre-Daniel Sarte, and Raymond Owens III. "Firm Fragmentation and Urban Patterns." *International Economic Review* 50 (2009): 143–86.

Roth, Alvin E. "Marketplaces, Markets, and Market Design." *American Economic Review* 108 (2018): 1609–58.

Rubin, Richard R., Mark Peyrot, and Christopher D. Saudek. "Effect of Diabetes Education on Self-Care, Metabolic Control, and Emotional Well-Being." *Diabetes Care* 12 (1989): 673–79.

Sabin, Paul. *The Bet: Paul Ehrlich, Julian Simon, and Our Gamble over Earth's Future*. New Haven: Yale University Press, 2013.

Saiz, Albert. "Immigration and Housing Rents in American Cities." *Journal of Urban Economics* 61 (2007): 345–71.

———. "Room in the Kitchen for the Melting Pot: Immigration and Rental Prices." *Review of Economics and Statistics* 85 (2003): 502–21.

San Diego Foundation. "San Diego's Changing Climate: A Regional Wake-Up Call." http://www.delmar.ca.us/DocumentCenter/View/1901/San-Diegos-Changing-Climate—-Regional-Wake-Up-Call_Summary-of-Focus-2050-Report_The-San-Diego-Found.

Savvides, Andreas, et al. "Chemical Priming of Plants against Multiple Abiotic Stresses: Mission Possible?" *Trends in Plant Science* 21 (2016): 329–40.

Scheffers, Brett R., et al. "Microhabitats Reduce Animal's Exposure to Climate Extremes." *Global Change Biology* 20 (2014): 495–503.

Schlenker, Wolfram, and Michael J. Roberts. "Nonlinear Temperature Effects Indicate Severe Damages to US Crop Yields under Climate Change." *PNAS* 106 (2009): 15594–98.

Schneider, Keith. "Chemical Plants Buy Up Neighbors for Safety Zone." *New York Times*, November 28, 1990.

Schwartz, John. "Canada's Outdoor Rinks Are Melting; So Is a Way of Life." *New York Times*, March 20, 2018.

Schwartz, John, and Richard Fausset. "North Carolina, Warned of Rising Seas, Chose to Favor Development." *New York Times*, September 18, 2018.

Severen, Christopher, Christopher Costello, and Olivier Deschênes. "A Forward-Looking Ricardian Approach: Do Land Markets Capitalize Climate Change Forecasts?" *Journal of Environmental Economics and Management* 89 (2018): 235–54.

Shelgikar, Anita Valanju, Patricia F. Anderson, and Marc R. Stephens. "Sleep Tracking, Wearable Technology, and Opportunities for Research and Clinical Care." *Chest* 150 (2016): 732–43.

Short, John R. "The West Is on Fire and the US Taxpayer Is Subsidizing It." *Conversation*, September 23, 2015. https://theconversation.com/the-west-is-on-fire-and-the-us-taxpayer-is-subsidizing-it-47900.

Shute, Nancy. "How an Economist Helped Patients Find the Right Kidney Donors." *NPR*, June 11, 2015. https://www.npr.org/sections/health-shots/2015/06/11/412224854/how-an-economist-helped-patients-find-the-right-kidney-donor.

Sieg, Holger, et al. "Estimating the General Equilibrium Benefits of Large Changes in Spatially Delineated Public Goods." *International Economic Review* 45 (2004): 1047–77.

Simon, Heather, et al. "Ozone Trends across the United States over a Period of Decreasing NOx and VOC Emissions." *Environmental Science and Technology* 49 (2014): 186–95.

Sinai, Todd, and Nicholas S. Souleles. "Owner-Occupied Housing as a Hedge against Rent Risk." *Quarterly Journal of Economics* 120 (2005): 763–89.

Singapore Government. Land Transport Authority. https://www.lta.gov.sg/content/ltaweb/en/roads-and-motoring/managing-traffic-and-congestion/electronic-road-pricing-erp.html.

Sjaastad, Larry A. "The Costs and Returns of Human Migration." *Journal of Political Economy* 70 (1962): 80–93.

Smith, James P., John J. McArdle, and Robert Willis. "Financial Decision Making and Cognition in a Family Context." *Economic Journal* 120 (2010): F363–80.

SMUD. "Rate Information: Low-Income Assistance and Nonprofit Discount." https://www.smud.org/en/Rate-Information/Low-income-and-nonprofits.

Solow, R. M. *Sustainability: An Economist's Perspective*. The Eighteenth J. Seward Johnson Lecture. Woods Hole, MA: Woods Hole Oceanographic Institution, 1991.

Sorrel, Charlie. "This Quaint English House Can Jack Itself Up on Stilts to Avoid Floods." *Fast Company*, July 21, 2016. https://www.fastcompany.com/3065783/this-quaint-english-house-can-jack-itself-up-on-stilts-to-avoid-floods.

State of California. Seismic Safety Commission. "Homeowner's Guide to Earthquake Safety." 2005 ed. https://ssc.ca.gov/forms_pubs/cssc_2005-01_hog.pdf.

St. John, Paige, and Anna M. Phillips. "Despite Fire after Fire, Paradise Continued to Boom—Until California's Worst Wildfire Hit." *Los Angeles Times*, November 13, 2018.

St. John, Paige, et al. "California Fire: What Started as a Tiny Brush Fire Became the State's Deadliest Wildfire." *Los Angeles Times*, November 18, 2018.

Stensrud, David J., et al. "Progress and Challenges with Warn-on-Forecast." *Atmospheric Research* 123 (2013): 2–16.

Stigler, George J. "The Cost of Subsistence." *Journal of Farm Economics* 27 (1945): 303–14.

Sun, Cong, Matthew E. Kahn, and Siqi Zheng. "Self-Protection Investment Exacerbates Air Pollution Exposure Inequality in Urban China." *Ecological Economics* 131 (2017): 468–74.

Sun, Menggi, and Leslie Scism. "Even after Last Year's Terrible Hurricanes, Insurers Are in Solid Shape." *Wall Street Journal,* June 30, 2018.

Sweet, Rod. "Why China Can Build High-Speed Rail So Cheaply." *Global Construction Review,* July 14, 2014. https://www.globalconstructionreview.com/sectors/why-china-can-build-high-speed-rail34socheaply7365/.

Swegal, Hayden. "The Rise and Fall of Almond Prices: Asia, Drought, and Consumer Preference." Bureau of Labor Statistics, *Beyond the Numbers* 6, no. 12 (2017).

Syverson, Chad. "What Determines Productivity?" *Journal of Economic Literature* 49 (2011): 326–65.

Tabuchi, Hiroko. "Tokyo Is Preparing for Floods 'Beyond Anything We've Seen.' " *New York Times,* October 6, 2017.

Tesla. "Gigafactory." https://www.tesla.com/gigafactory.

Thaler, Richard H. *Misbehaving: The Making of Behavioral Economics.* New York: W. W. Norton, 2015.

Tiebout, Charles M. "A Pure Theory of Local Expenditures." *Journal of Political Economy* 64 (1956): 416–24.

Trafton, Anne. "Comparing Apples and Oranges." MIT News, Massachusetts Institute of Technology, April 30, 2012. news.mit.edu/2012/fruit-spoilage-sensor-0430.

Troesken, Werner. "Race, Disease, and the Provision of Water in American Cities, 1889–1921." *Journal of Economic History* 61 (2001): 750–76.

———. *Water, Race, and Disease.* Cambridge, MA: MIT Press, 2004.

University of Pennsylvania. Wharton Risk Center. "Portland Flood Insurance Study." https://riskcenter.wharton.upenn.edu/incubator/upgrading-flood-insurance/portland-flood-insurance-study/.

US Army Corps of Engineers. New York District Website. "NY & NJ Harbor & Tributaries Focus Area Feasibility Study (HATS)." https://www.nan.usace.army.mil/Missions/Civil-Works/Projects-in-New-York/New-York-New-Jersey-Harbor-Tributaries-Focus-Area-Feasibility-Study/.

US Department of Agriculture. Economic Research Service. "Food Expenditure Series." https://www.ers.usda.gov/data-products/food-expenditures/food-expenditures/#Food%20Expenditures.

———. Farm Credit Administration. "Crop Insurance Covers Most Major Crops." September 28, 2017. https://www.fca.gov/template-fca/download/Economic Reports/CropInsuranceCoversMostMajorCrops.pdf.

———. National Agricultural Statistics Service. "Farms and Farmland Numbers, Acreage, Ownership, and Use." September 2014. www.nass.usda.gov/Publications/Highlights/2014/Highlights_Farms_and_Farmland.pdf.

US Energy Information Administration. "Air Conditioning in Nearly 100 Million U.S. Homes." RECS 2009, August 19, 2011. https://www.eia.gov/consumption/residential/reports/2009/air-conditioning.php.

———. "Household Energy Use in Arizona." 2009. https://www.eia.gov/consump
tion/residential/reports/2009/state_briefs/pdf/AZ.pdf.

US Environmental Protection Agency. "DC Water's Environmental Impact Bond: A
First of Its Kind." April 2017. https://www.epa.gov/sites/production/
files/2017-04/documents/dc_waters_environmental_impact_bond_a_first_of_
its_kind_final2.pdf.

———. "Flood Resilience: A Basic Guide for Water and Wastewater Utilities."
August 2015. https://www.epa.gov/sites/production/files/2015-08/docu
ments/flood_resilience_guide.pdf.

US House Committee on Science, Space, and Technology. Hearing: "The Future of
Forecasting: Building a Stronger U.S. Weather Enterprise." Testimony of Rich
Sorkin, Chief Executive Officer of Jupiter Intelligence, May 16, 2019. https://
science.house.gov/imo/media/doc/Sorkin%20Testimony.pdf.

USC Viterbi. Department of Computer Science. "Data Science." https://www.
cs.usc.edu/academic-programs/masters/data-science/.

Velazco, Chris. "Arable's Mark Crop Sensors Give Farmers a Data-Driven Edge."
Engadget, January 12, 2018. https://www.engadget.com/2018/01/10/arables-
mark-crop-sensors-give-farmers-a-data-driven-edge/.

Voelker, Rebecca. "Vulnerability to Pandemic Flu Could Be Greater Today Than a
Century Ago." *JAMA* 320 (2018): 1523–25.

Wagner, Gernot, and Martin L. Weitzman. *Climate Shock: The Economic Consequences of a Hotter Planet.* Princeton, NJ: Princeton University Press, 2016.

Waldfogel, Joel. "The Median Voter and the Median Consumer: Local Private Goods
and Population Composition." *Journal of Urban Economics* 63 (2008): 567–82.

Wallace-Wells, David. *The Uninhabitable Earth: Life after Warming.* New York: Tim
Duggan Books, 2019.

Walmart. "Enhancing Resilience in the Face of Disasters." https://corporate.
walmart.com/2017grr/community/enhancing-resilience-in-the-face-of-disasters.

Walters, Joanna. "Plight of Phoenix: How Long Can the World's Least Sustainable
City Survive?" *Guardian,* March 20, 2018.

Wang, Jin. "The Economic Impact of Special Economic Zones: Evidence from Chinese Municipalities." *Journal of Development Economics* 101 (2013): 133–47.

Weitzman, Martin L. "On Modeling and Interpreting the Economics of Catastrophic Climate Change." *Review of Economics and Statistics* 91 (2009): 1–9.

———. "What Is the 'Damages Function' for Global Warming—And What Difference Might It Make?" *Climate Change Economics* 1 (2010): 57–69.

Wibbenmeyer, Matthew, Sarah E. Anderson, and Andrew J. Plantinga. "Salience
and the Government Provision of Public Goods." *Economic Inquiry* 57 (2019):
1547–67.

Wolak, Frank A. "Do Residential Customers Respond to Hourly Prices? Evidence
from a Dynamic Pricing Experiment." *American Economic Review* 101 (2011):
83–87.

Wolfram, Catherine, Orie Shelef, and Paul Gertler. "How Will Energy Demand Develop in the Developing World?" *Journal of Economic Perspectives* 26 (2012): 119–38.

Wood, Matt. "World War I-Era Maps Help Track History of Kelp Forests in the Pacific Northwest." *Forefront,* December 20, 2017. https://www.uchicagomedi cine.org/forefront/biological-sciences-articles/world-war-i-era-maps-help-track-history-of-kelp-forests-in-pacific-northwest.

World Bank. "World Bank Catastrophe Bond Transaction Insures the Republic of Philippines against Natural Disaster-Related Losses up to US$225 million." Press Release, November 25, 2019. https://www.worldbank.org/en/news/press-re-lease/2019/11/25/world-bank-catastrophe-bond-transaction-insures-the-republic-of-philippines-against-natural-disaster-related-losses-up-to-usd225-million.

"The World Is Losing the War against Climate Change." *Economist,* August 2, 2018. https://www.economist.com/leaders/2018/08/02/the-world-is-losing-the-war-against-climate-change.

Wuebbles, D. J., et al. "Executive Summary." In *Climate Science Special Report: Fourth National Climate Assessment,* ed. D. J. Wuebbles et al., 1:12–34. US Global Change Research Program, 2017. doi: 10.7930/J0DJ5CTG.

Yan, Wu. "Chinese Team Succeeds in Planting Saltwater Rice in Dubai's Desert." *China Daily,* May 31, 2018. www.chinadaily.com.cn/a/201805/31/WS5b0fb-51fa31001b82571d787.html.

Yeginsu, Ceylan. "KFC Will Test Vegetarian 'Fried Chicken,' Original Herbs and Spices Included." *New York Times,* June 8, 2018.

Zheng, Siqi, and Matthew E. Kahn. "A New Era of Pollution Progress in Urban China?" *Journal of Economic Perspectives* 31 (2017): 71–92.

Zheng, Siqi, et al. "Air Pollution Lowers Chinese Urbanites' Expressed Happiness on Social Media." *Nature Human Behaviour* 3 (2019): 237–43.

Zhu, Chunwu, et al. "Carbon Dioxide (CO_2) Levels This Century Will Alter the Protein, Micronutrients, and Vitamin Content of Rice Grains with Potential Health Consequences for the Poorest Rice-Dependent Countries." *Science Advances* 4 (2018): eaaq1012.

Zillow. "Home Values." https://www.zillow.com/mi/home-values/.

———. "Paradise, California, Home Values." https://www.zillow.com/paradise-ca/home-values/.

Zivin, Joshua Graff, and Matthew E. Kahn. "Industrial Productivity in a Hotter World: The Aggregate Implications of Heterogeneous Firm Investment in Air Conditioning." NBER Working Paper No. 22962, December 2016.

Acknowledgments

A highlight of the annual Allied Social Sciences Association meeting is the opportunity to visit the exhibition hall. Each academic press displays its new books, and the editors attend and meet with potential book writers. At the January 2018 meeting in Philadelphia, I discussed my nascent climate change adaptation book project with Seth Ditchik, editorial director of Yale University Press.

Amid the covid-19 pandemic, climate change receives little current media attention, but this challenge lurks and affects almost every sector of our lives. Over the past decade, environmental economists have made great progress studying how weather events ranging from extreme heat to drought to sea level rise affect our economy, our health, and our quality of life. Given that climate change poses the risk of exacerbating such weather shocks, it is crucial to have a better understanding of how recent weather events have affected our economy. Many environmental economists are now using big data to generate fascinating facts about these impacts.

I talked to Seth about my goal of writing an accessible book that highlights the key ideas and findings in this emerging literature. My book's core thesis is that we are becoming ever better at adapting to weather shocks. Seth pushed me to explain why and to examine what new rules of the game would accelerate this adaptation progress.

Seth encouraged me to read Nobel Laureate Robert Shiller's 2004 book *The New Financial Order: Risk in the 21st Century*. This brilliant work presents many new ideas on how to reduce our exposure to risks ranging from unemployment to declining home and asset values. Shiller argues that new markets and new regulations could be introduced to reduce our overall risk exposure. Given that climate change poses many new risks, I immediately saw how I could use the structure of Shiller's book to improve my book's arc. In the first half of this book, I explore the microeconomics of how climate change affects key sectors of the economy. In the second half, I discuss how new markets and regulatory reforms together can accelerate the pace of climate change adaptation. Recognizing the challenge of changing policies, I carefully discuss the relevant interest group politics.

In emphasizing the key role that human capital and markets play in facilitating adaptation, I return to a key theme of both Julian Simon and the PERC think-tank. During the summers of 2016 and 2017, I served as a Julian Simon Fellow at PERC. My discussions with Terry Anderson, Reed Watson, and PERC seminar participants have sharpened my thinking. I thank the Searle Freedom Trust Foundation for generously funding this project.

In writing this book, I have benefited greatly from comments received from my editor, Seth Ditchik, manuscript editor Laura Jones Dooley, and Carol Kahn, Mac McComas, and Steve Olson. Thanks to my many coauthors whose work I discuss throughout this book. Special thanks to Brian Casey and Nolan Jones, my coauthors on chapter 10. Over the years, I have taught many talented undergraduates environmental economics. Brian and Nolan are two of my star students, and they have taught me plenty.

Index

adaptation, 1–17; accelerating progress in, 14–17; agricultural production and, 203–4; animals' survival and, 239–41; big data and, 157–61, 163–66, 243; coping with risks, 2; great race for, 17; at household level, 10–12; human capital and, 237–38; hypothesis, 8, 28, 76, 243; microeconomic perspective and, 5–9; mitigation challenge and, 2–5; poverty and, 241–42; productivity and, 12–13; public infrastructure and, 12; real estate markets and, 14; science and economics, integration of, 9–10; social cost reduction via, 46. *See also* laws and regulations for adaptation

Africa: agricultural production in, 213, 215, 217; civil wars in, 26–27; urbanization in, 73

African Risk Capacity, 213, 217

agricultural sector, 16, 201–22; adaptation transition paths, 203–4; big data and, 211–14; education for new farmers, 215–16; farmer variation and, 204–6; financial market access and, 216–18; food consumption and, 55–56; GMOs and, 208–11; high heat

and, 202; indoor farming, 206–8; international trade and, 229–31; private and public investment in research and, 220–22; public policies for, 218–20; urbanization and climate shocks, 74; water consumption and, 153, 209

Agriculture Department (USDA), 201

Airbnb, 156

air conditioning: declining price of, 76; demand for, 9; feedback loops of emissions and, 48–49; health and mortality rates, 47; poverty and, 9, 67–69, 165–66; real estate markets and, 135–36; in schools, test scores and, 7–8; unanticipated heat waves and, 158; work environment and, 107–8

air pollution: air filters and masks for, 44, 51–52, 149; apps for, 147–49; child development and, 62; fires and, 2, 51, 147; mitigating, 43–45; monitoring, 164; quality of life and, 51–52; race and exposure to, 69; wildfires and, 2, 51, 147. *See also* automobiles

Akerlof, George, 194